Raj Jani

MRP II

Chapman & Hall
Materials Management/Logistics Series

Eugene L. Magad, Series Editor

William Rainey Harper College

Distribution: Planning and Control, by David F. Ross

Total Materials Management: Achieving Maximum Profits through Materials/Logistics Operations, Second Edition by Eugene L. Magad and John M. Amos

International Logistics, by Donald Wood, Anthony Barone, Paul Murphy & Daniel Wardlow

Global Purchasing: Reaching for the World, by Victor Pooler

Practical Handbook of Warehousing, Third Edition, by Kenneth B. Ackerman

Practical Handbook of Industrial Traffic Management, Seventh Edition, by Leon Wm. Morse

MRP II, by John W. Toomey

MRP II

Planning for Manufacturing Excellence

John W. Toomey
Principal, John Toomey & Associates, Arlington Heights Ilinois

An International Thomson Publishing Company
New York • Albany • Bonn • Boston • Cincinnati • Detroit • London • Madrid • Melbourne
Mexico City • Pacific Grove • Paris • San Francisco • Singapore • Tokyo • Toronto • Washington

Printed in the United States of America

For more information contact:

Chapman & Hall
115 Fifth Avenue
New York, NY 10003

Chapman & Hall
2-6 Boundary Row
London SE1 8HN
England

Thomas Nelson Australia
102 Dodds Street
South Melbourne, 3205
Victoria, Australia

Chapman & Hall GmbH
Postfach 100 263
D-69442 Weinheim
Germany

Nelson Canada
1120 Birchmount Road
Scarborough, Ontario
Canada M1K 5G4

International Thomson Publishing Asia
221 Henderson Road #05-10
Henderson Building
Singapore 0315

International Thomson Editores
Campos Eliseos 385, Piso 7
Col. Polanco
11560 Mexico D.F.
Mexico

International Thomson Publishing - Japan
Hirakawacho-cho Kyowa Building, 3F
1-2-1 Hirakawacho-cho
Chiyoda-ku, 102 Tokyo
Japan

1 2 3 4 5 6 7 8 9 XXX 01 00 99 98 97 96

Library of Congress Cataloging-in-Publication Data

Toomey, John W.
MRP II: planning for manufacturing excellence / John W. Toomey.
P. Cm.
Includes bibliographical references and index.
ISBN 0-412-06581-9 (alk. paper)
1. Manufacturing resource planning. I. Title.
TS176.T633 1996 95-46335
658.5--dc20 CIP

Visit Chapman & Hall on the Internet http://www.chaphall.com/chaphall.html

To order this or any other Chapman & Hall book, please contact **International Thomson Publishing, 7625 Empire Drive, Florence, KY 41042.** Phone (606) 525-6600 or 1-800-842-3636. Fax: (606) 525-7778. E-mail: order@chaphall.com.

For a complete listing of Chapman & Hall titles, send your request to **Chapman & Hall, Dept. BC, 115 Fifth Avenue, New York, NY 10003.**

Preface

After attending the International American Production and Inventory Control Conference in 1972, I became part of the "MRP Crusade." For the next 2 years, I directed a program at my company to install material requirements planning, and following that, an additional 2 years was devoted to the implementation of capacity requirements planning. The results, although successful, were not good enough to qualify us as a "Class A" user, nor did our manufacturing performance meet the present-day requirements of a world-class manufacturer.

In the intervening years, our understanding of the role of planning and its relationship to manufacturing execution has expanded and clarified. Competition from abroad as well as domestic competitors has hastened the increased body of knowledge necessary for efficient manufacturing. The relationship of the manufacturing system to the increased emphasis on concepts such as improved customer service, continuous improvement, and employee empowerment must be considered.

This book is an organized collection of material I have used in teaching at community colleges and management seminars. It is an extension of the basic logic of MRP in Joseph Orlicky's *Material Requirements Planning* (1975), coupled with the increased understanding of materials management requirements as well as the availability of more sophisticated software. This book will explain the principles of MRP and how they can be utilized in distribution, job shop, and repetitive and process manufacturing environments.

The object of the book is to meet the training need of MRP courses offered at 4-year colleges, community colleges, and company in-house programs. It is also directed to material and manufacturing practitioners who require in-depth knowledge of MRP or who would use it as a reference book to assist them in APICS certification.

The material is organized to first present the perspective of materials management (Chapters 1 and 2). Defining the product and process is then specified (Chapter 3). The planning functions for demand management, MRP, capacity planning, and distribution planning follow (Chapters 4–8). Executing the plan for job shop, process, and just-in-time operations is then detailed (Chapters 9–11). Finally, system implementation including organization and measurements are discussed (Chapter 12). Simple case studies and solutions are presented in each chapter with the exception of Chapter 1. Each chapter concludes with a 10-question quiz. The answers to the quizzes can be found in Appendix C.

I have learned from this effort that writing a book is no easy task. Jim Nolan describes the experience as similar to writing a 400-page homework assignment. I would like to acknowledge the assistance and encouragement from Eugene Magad of Harper College as well as the benefits received from manuscript reviews by Jack Gips of Jack Gips Inc., Gary Midkiff of Friedman & Associates, Tom Setlik of Tempel Steel Company, and Larry Sutherland of System Software Associates. It is amazing how critical points can be overlooked or, worse, misstated due to the wrong word in the right place. The above reviewers have been invaluable in their suggestions and corrections.

Finally, I would like to thank the personnel at Letter Perfect Secretarial Service, Inc., who had to put up with my penmanship, and my wife Joan, who had to put up with me.

John W. Toomey

Contents

MRP II

1
Introduction

In the decade of the nineties, MRP II—Manufacturing Resource Planning has continued to be the predominant system for manufacturing control. It is a key ingredient of world-class manufacturing which calls for

Improved customer service

Elimination of waste

Continuous improvement

IMPROVED CUSTOMER SERVICE

In the past, customer service was measured by such factors as percent shipped on time or average days late on past due orders. The more modern and broader definition of customer service is the ability of the company to meet the total needs of the customer. The goal is continuous customer satisfaction not only with the product but also with the company. This can be accomplished not only by meeting the basic requirements of a product with adequate quality, on-time delivery, and reasonable pricing but also by working with the customer to assist in meeting the marketplace demands of the customer.

Quality is more than a defect-free product that meets specific physical specifications. Quality is also a product designed to give the customer reliable service and fitness for use. The definition can be expanded to include the manner in which the product is packaged and the clarity of included manuals or instructions. Customer communications such as delivery promises, inquiry responses, and invoicing are also quality considerations.

Delivery performance is more than shipping on time. It is understood that the immediate availability of a stocked item (off the shelf) must be met. The delivery of a made-to-order product will best meet the needs of the customer with reduced manufacturing lead time. The shorter the cumulative lead time, the easier the forecasting task for both the customer and the manufacturer.

Flexibility in the manufacturing process will allow quick response when the customer is in need of reduced lead time.

ELIMINATION OF WASTE

Waste is any cost, direct or indirect, that does not add to the value of the product. Traditionally, waste has been considered to be quality losses such as scrap or rework. Often the measurement might show a scrap rate of 4% and a rework cost of 2%. In this circumstance, the quality cost would seem to be 6%. If closer inspection of quality-related costs such as engineering change orders, technical time spent, customer service time, additional freight, and inspection were considered, the initial 6% cost of quality loss might well grow to 20%.

Quality costs are not the only non-value-added costs that are considered waste. Just as inspection adds nothing to the value of the product, material handling and other indirect labor costs are circumspect. Control activities such as transaction reporting are being audited as they, too, add no value to the manufactured product. Employees may be fully occupied, but if their work assignments are not utilizing their abilities, there is a waste of intelligence which perhaps is the most important asset.

Money invested in inventory, although listed on balance sheets as assets, is, in reality, wasted dollars that could be better spent in product or process development. Excessive work-in-process not only adds no value to the product and holds dollars captive, it increases manufacturing lead time and reduces shop-floor flexibility while requiring additional storage space.

CONTINUOUS IMPROVEMENT

Continuous improvement is a strategy that calls for a never-ending quest to eliminate waste. Whereas some projects such as a new system implementation or a capital equipment expansion have a planned completion date, continuous improvement is a lifelong journey. This is because there is no end to improvement.

Improvement goes beyond immediate problem solving. There must be a determination of what will differentiate the company from the competitor. To accomplish this, a management system based on the company as a whole, rather than on a departmental basis, must answer the question, "What is needed and what should be done?" Proper performance measures must be an integral part of the process.

These goals are best achieved with planning systems which allow for optimum manufacturing execution.

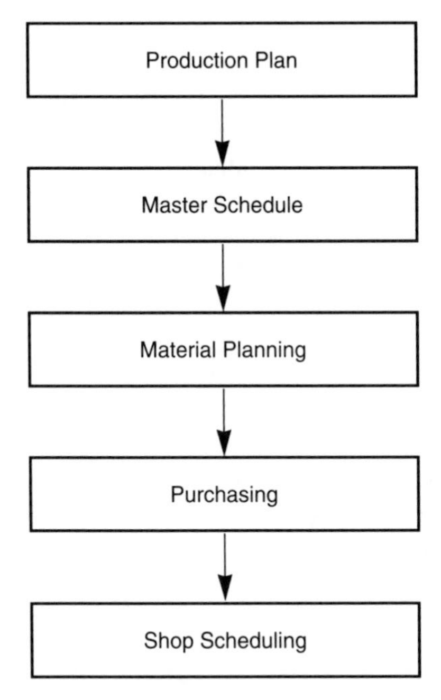

Figure 1-1. Material requirements planning—MRP.

DEFINITIONS

The term "MRP" has three definitions that have evolved since the initial introduction in the 1960s. These definitions as listed in the *APICS Dictionary* are as follows:

> **Material Requirements Planning (MRP)**—A set of techniques as shown in Figure 1-1 that uses bills of material, inventory data, and the master production schedule to calculate requirements for materials. It makes recommendations to release replenishment orders for material. Further, because it is time phased, it makes recommendations to reschedule open orders when due dates and need dates are not in phase. Time-phased MRP begins with the items listed on the Master Production Schedule and determines (1) the quantity of components and materials required to fabricate those items and (2) the date that the components and material are required. Time-phased MRP is accomplished by exploding the bill of materials, adjusting for inventory quantities on hand or on order and offsetting the net requirements by the appropriate lead times.

Closed-Loop MRP—A system built around material requirements planning that includes the additional planning functions of sales and operations (production planning, master production scheduling, and capacity requirements planning). Once this planning phase is complete and the plans have been accepted as realistic and attainable, the execution functions come into play. These include the manufacturing control functions of input–output (capacity) measurement, detailed scheduling and dispatching, as well as anticipated delay reports from both the plant and suppliers, supplier scheduling, and so on. The term "closed loop" implies that not only is each of these elements included in the overall system but also that feedback is provided by the execution functions so that the planning can be kept valid at all times. Closed-loop MRP is illustrated in Figure 1-2.

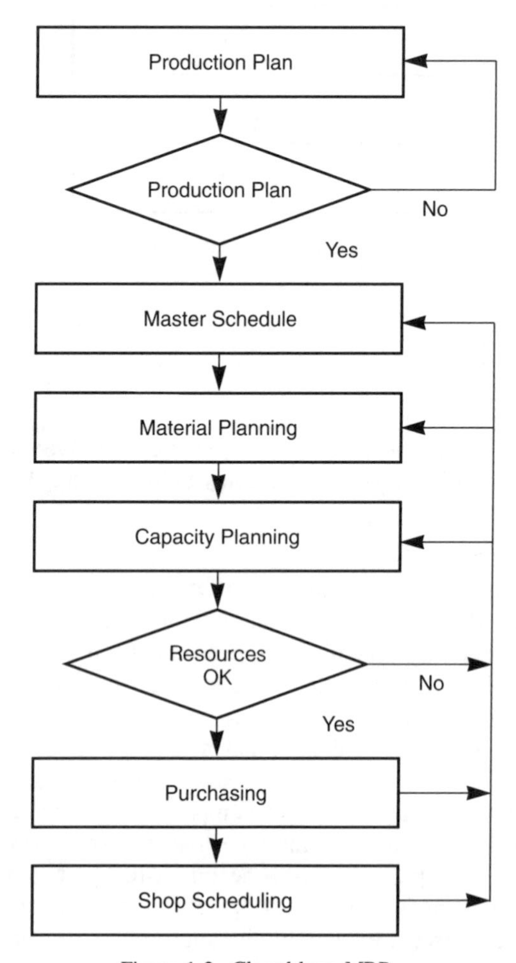

Figure 1-2. Closed-loop MRP.

Manufacturing Resource Planning (MRP II)—A method for the effective planning of all resources of a manufacturing company. Ideally, it addresses operational planning in units and financial planning in dollars and has a simulation capability to answer "what-if" questions. It is made up of a variety of functions, each linked together: business planning, sales and operations (production planning), master production scheduling, material requirements planning, capacity requirements planning, and the execution support systems for capacity and material. Output from these systems is integrated with financial reports such as business plan, purchase commitment report, shipping budget, inventory projections in dollars, and so on. Manufacturing resource planning is a direct outgrowth and extension of closed-loop MRP as shown in Figure 1-3.

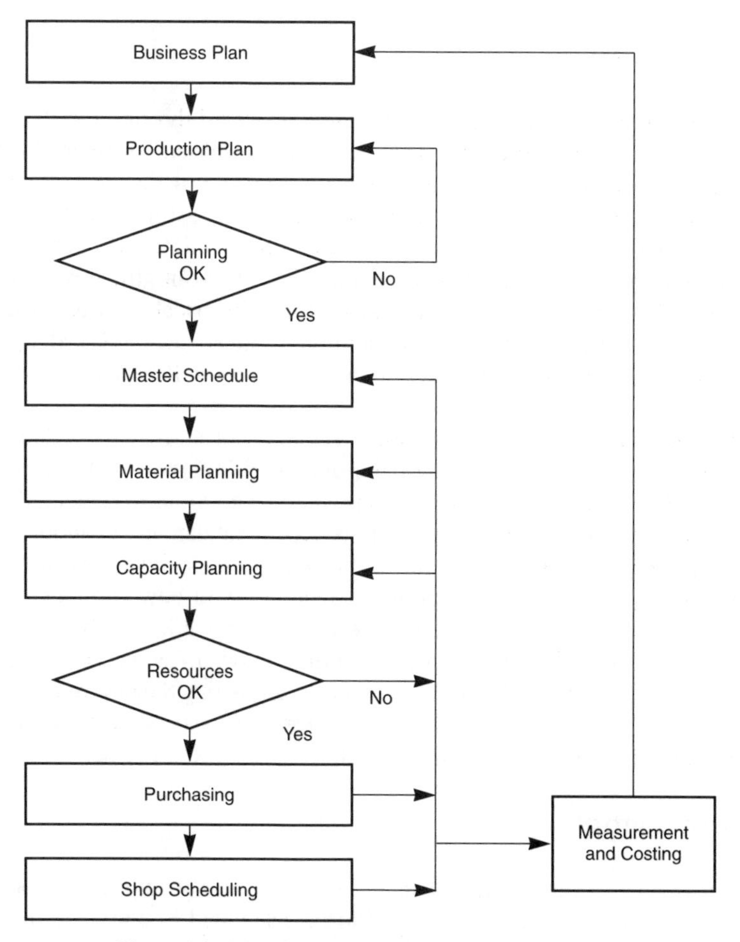

Figure 1-3. Manufacturing resource planning—MRP II.

As the reader progresses through this book, the definitions above will be studied in detail.

WHERE WE HAVE BEEN

To better understand the role MRP plays in industry, we should look back at the course of actions that have evolved in the past in an effort to best manage the manufacturing processes. One of the first scientific techniques put to use was the "Economic Lot Size" formula developed in 1913 and which balances the cost of carrying inventory with the cost of setting up the production run. The EOQ (Economic Order Quantity) is easily calculated based on four variables and has been one of the most used (and abused) techniques ever since.

When to Order

The question of when to order was initially answered through reorder points maintained manually on cards. Reorder points are based on the lead time to manufacture or procure the item and the usage rate of the item during the lead time. When the usage rate forecast is based on past history, the process calls for the calculation of the "statistical order point." The problem of maintaining historical data in precomputer times was simplified with the use of a technique called exponential smoothing. Historical data are used not only in forecasting but also in the calculation of safety stocks. Reorder points, forecasting, lot sizing, and safety stock techniques will be covered later in the book.

Shortcomings in reorder point logic such as not allowing for dependent demand of one item to another or the unrelated lot sizes of the dependent relationships were understood for many years. The logic for calculating the net requirements on one item based on its dependent demand to another item, the time phasing of the item, and projected available inventory were also known for as many years, but until the availability of new computers in the 1960s, the required data processing was unmanageable. With the storage capacities and speed of the third-generation computers as well as a software concept called the Bill of Material processor, calculating the new requirements of all manufactured and purchased items by time period became practical. This set of techniques became known as MRP—Material Requirements Planning.

How to Control

At any point in time, a typical manufacturing facility might be producing 1500 items with 7000 operations still to be scheduled to complete the 1500 items. Using reorder point control, these items would have been released to the shop based on past average usage and expected manufacturing lead time. There would be no understanding of future dependent requirements nor the time phasing of those demands. At the time of order release, the due date may have been based on a 10-week lead time, but during the 10 weeks of run, there would be no knowledge of changes in demand that would call for the item to complete in week 8 rather than week 10. The required expediting would not take place until an actual shortage occurred. On the other hand, there could be demand change that would call for required delivery not in week 10, but week 13. Again, there was no knowledge that the item could be moved back 3 weeks (deexpediting).

With MRP control, the same 1500 items with the same 7000 operations to completion might well remain, as the lot sizes and the lead times have not changed. What would change would be an understanding of each item's planned dependent demand and the time period(s) of the demand. The valid due date of the item demand as well as the scheduled due date at time of order release are known and compared. If the requirement is called for in week 8 compared to the scheduled due date of week 10, expediting (moving up the schedule) is recommended. Again, if the required due date is week 13, compared to a schedule due date of week 10, deexpediting (moving back the schedule) is recommended.

Assuming both accurate inventories and bills of material, an MRP system will give valid due dates of all manufactured and purchased items required for planned end items. What has been overlooked by many practitioners is the difference between a valid due date and a realistic due date. Due to unforeseen usage or scrap, an item originally planned for delivery in 10 weeks may be required in 2 weeks. The week 2 due date is valid, but expected delivery in week 2 may not be at all realistic. An MRP system will supply *valid* due dates, but the operating system can only execute to a *realistic* due date. The distinction between the two has not always been understood.

MRP to Just-In-Time

Led by the "MRP Crusade" of the American Production and Inventory Control Society, the use of MRP increased at an incredible rate in the 1970s. Not only did the scope expand to closed-loop and MRP II, but the available software expanded in numbers and capability. What started with mainframe-based systems moved to mid-size in the 1980s and then to microprocessors. In spite of MRP growth in system sophistication, user knowledge, and successful implementations, manufacturing in the United States in the 1980s was not always competitive on a world-class basis, especially when compared to Japan. To become more competitive, many companies have turned to Just-In-Time (JIT) philosophy of manufacturing.

Some erroneously thought that with Just-In-Time inventory, MRP manufacturing systems would no longer be needed. After over 10 years experience with JIT concepts, it has been found that customers do not always order in continuous steady patterns, that setups and lot sizes cannot be easily reduced as much as desired, that some suppliers cannot or will not deliver based on reasonable flow rates, and that the diverse work force is not easily trained to be multifunctional operators and contributing members of problem-solving work teams. In spite of these problems, we realize that in world competition, there must be continuing efforts to improve our manufacturing operations but not necessarily in the dynamic high-impact improvement projects common in the past but in slow continuous improvement steps as practiced by the Japanese.

WHERE WE MUST GO

The MRP crusade of 20 years ago turned out to be a disappointment to many. Consultants have had clients advise them to use another term rather than MRP when discussing manufacturing planning systems to top management and operating people. The "failure" of MRP was not due to poor software or poor systems implementation, but to the false expectation that MRP could execute the manufacturing plan in an environment of uncontrolled processes. As we take a second look for the "best" planning system, the logic of MRP as stated in Orlicky's *Material Requirements Planning,* in 1975, cannot be challenged. The material requirements of that time, MRP, was an inventory-planning system that listed what parts, both manufacturing and purchased, were required to meet a master plan.

Evolution by Acronym

MRP is driven by the Master Plan or Master Production Schedule (MPS) which represents what the company plans to produce expressed in specific configurations, quantities, and dates. Because MRP assumes infinite capacity, the master plan must be realistic. "Closed-loop MRP" was developed to review capacity in order to allow for adjustments to the master plan to make that plan attainable. Closed-loop MRP uses the logic of MRP as well as detailed routings and capacities found in the manufacturing data base. MRP II (Manufacturing Resource Planning) is an extension of closed-loop MRP and includes financial planning and "simulation" capabilities. Again, the logic of MRP is the base of this system which not only assists in realistic master planning but is an instrument of Business Planning. Whereas MRP, MRP II, and closed-loop MRP are definitive one from the other, actual practice finds the terms used interchangeably.

The logic of MRP has been extended from manufacturing systems to distribution systems. DRP (Distribution Resource Planning) is a tool that plans key resources in distribution and is integrated with the manufacturing plan. Whereas the Master Production Schedule (the parent) drives the MRP, the anticipated needs of the distribution system drive the Master Production Schedule. With DRP, the warehouse is the parent and the Master Production Schedule becomes the component.

To make an MRP system operate effectively, certain disciplines are required, such as inventory record accuracy, data-base integrity, employee understanding, realistic master plans, and so forth. When the system is operating properly, the result will be valid and realistic plans. The plans are worthless if they are not executed as planned. World-class manufacturing cannot be achieved without a first-class plan coupled with world-class execution.

MRP AND MANUFACTURING EXECUTION

MRP was originally designed to control operations in job-shop production environments. Job-shop plants are organized with functional departments such as saw, mill, grinding, and so on. Production lots are routed from department to department and controlled by shop orders (work orders, manufacturing orders, etc.). Job-shop production often results in long lead times, large work-in-process inventories, and lack of responsiveness to requirements. The Just-In-Time philosophy calls for short lead times, reduced work-in-process, and flexibility to demand by producing in a repetitive process environment.

Just-In-Time concepts such as setup reductions and product and process simplification have made great strides in the effort to achieve synchronous product flow but often a specific resource becomes a bottleneck in the process flow. The theory of constraints philosophy accepts the reality of the bottleneck and advocates close management of the bottleneck, which, in turn, will pace the entire system. The process focuses on identifying and exploiting the constraint (bottleneck). Buffer inventory ahead of nonbottleneck operations are planned to allow for temporary bottlenecks due to statistical fluctuations in the process. If the bottleneck is broken, the process of identifying a new constraint is started over.

The reality of both products and process calls for execution concepts ranging from continuous synchronized flow through dedicated equipment to nonsynchronized interrupted flow through functional manufacturing departments. Whatever the execution concepts, the operation must be based on solid plans generated by an MRP plan modified to the needs of the process.

BENEFITS OF MRP II

The benefits of a well-run MRP II program are numerous. Number one is improved customer service which can lead to increased sales or, in increasing competitive situations, the ability to maintain existing sales levels. Improved service can be brought about by faster response to customer needs through reduced lead times. Meeting schedules and, therefore, on-time shipments will allow "promises kept."

Organized plans that are attainable for both in-house manufacturing and purchasing will ultimately increase productivity and therefore reduce costs of both manufactured and purchased items. Reductions in the cost of dollar investment and space will be brought about by lower levels of raw material, in-process, and finished goods inventories.

In the long run, the greatest benefit, not easily measured, will be the benefit of everyone in the operation "singing out of the same hymnal." MRP II starts with a business plan and ultimately will involve every function within the organization. This will result not only in a reduction of waste but also in a reduction in finger pointing.

This book will cover the concepts, requirements, and logic of manufacturing and distribution planning and their place in the continuous manufacturing revolution of the nineties. The goal will be to assist the reader in understanding the basics of planning and planning's relationship to intermittent (job shop), repetitive, and process flow execution. The basics discussed will be the building blocks of knowledge required to properly define a desired planning system. It will not be a book of computer hardware or software but will help the reader to be a better evaluator of a system's computer requirements once the system has been defined.

QUIZ

1. Modern methods of inventory management were made possible with the introduction of
   I. Kardex files
   II. machine tools
   III. computers
   IV. statistics

   a. I　　c. III
   b. II　　d. IV

2. Service level is
   a. part of the bill of material
   b. a measure of delivery performance

c. unique part number
d. a criteria of MRP

3. Customer service is only measured by on-time delivery.
   a. True
   b. False

4. MRP II refers to
   a. material requirements planning
   b. manufacturing resource planning
   c. closed-loop MRP
   d. time-phased order point

5. The economic lot size formula (EOQ) balances the cost of carrying the inventory with
   I. the reorder point
   II. the cost of setting up the run
   III. the selling of the product
   IV. the lead time

   a. I and IV
   b. II
   c. I and III
   d. II and III

6. MRP is designed for dealing with
   I. dependent demand
   II. independent demand
   III. discontinuous service
   IV. Nonuniform demand

   a. I and II
   b. III and IV
   c. I, III, and IV
   d. II, III, and IV

7. Which of the following applications would be best suited to an order point systems?
   a. A product manufactured to customer order
   b. A subassembly that is used in one finished-goods item
   c. Component part inventory having a lumpy demand pattern
   d. A finished-goods item sold off the shelf with level usage

8. MRP is a system for
   I. material requirements
   II. rescheduling recommendations
   III. Execution

a. I
b. I and II
c. I and III
d. I, II, and III

9. MRP planning is required for
   I. job-shop operations
   II. Just-In-Time facilities
   III. operations with bottlenecks
   IV. nonsynchronized flow operations

   a. I and II
   b. I and III
   c. II and IV
   d. All of the above

10. Implementation of an MRP system guarantees a successful manufacturing operation.
    a. True
    b. False

BIBLIOGRAPHY

APICS Dictionary (7th ed.). Falls Church, VA: American Production and Inventory Control Society, 1992.

Fogarty, D. W., Blackstone, J. H., Jr., and Hoffman, T. R., *Production and Inventory Management*. Cincinnati: South-Western Publishing, 1991.

Goldratt, E. M., *The Haystack Syndrome*. Croton-on-Hudson, NY: North River Press, 1990.

Hall, R. W., *Zero Inventories*. Homewood, IL: Dow Jones-Irwin, 1983.

Orlicky, J., *Material Requirements Planning,* New York: McGraw-Hill, 1975.

Wight, O. W., *MRP II: Unlocking American Productivity Potential*. Boston: CBI Publishing, 1982.

2
The Role of Inventory Management

The role of inventory management is to maintain a desired stock level of specific products or items. The desired level is a function of customer service requirements and the cost of inventory investment. Once the parameters are determined, the challenge is how much to order, when to order, and how to control ongoing activities. The mission is to address the activities and techniques to best manage inventories. The determination of desired inventory levels requires the consideration of the attributes of each item.

THE ATTRIBUTES

Usage

The nature of the usage of an item may be as a uniform (continuous) rate or an intermittent (lumpy) rate. Factors controlling the usage are the marketplace or manufacturing requirements. Inventory for distribution use (the marketplace) as compared to manufacturing use are driven by forecasted demand or customer orders. These finished goods items may have been manufactured products such as cars by General Motors or purchased products such as shirts by Wal-Mart. Distribution inventory is customer oriented, whereas manufacturing inventory is product oriented. The goal of distribution inventory is to serve the customer, whereas the goal of manufacturing inventory is for that inventory to be completely converted to the next level of the product structure. The ultimate goal of the manufacturing system is to coordinate the production rate with the sales rate.

Cost

Inventory must be evaluated. If a single item is manufactured, the task is simple. Actual costs of the material, labor, and overhead are collected, and actual unit cost is calculated by dividing the total cost by the number of units produced. The inventory is then valued by multiplying the number of items in stock by the actual unit cost. Actual cost systems are not practical when multiple items are produced and overhead costs, not directly related to specific items, must be allocated.

When the material, labor, and overhead cost is estimated based on anticipated product mix and levels of operation, the resultant estimate is the unit standard cost. The standard unit cost is used for the following:

1. Inventory valuation
2. Selling price consideration
3. Performance measurement through comparison of actual to standard cost (cost variances)

Cost allocation can be most difficult, and if the limitation and the implications of standard cost systems are not understood, erroneous strategic conclusions can be made. Other cost systems such as direct costing and activity-based costing (ABC) may be more effective in performance measurement and decision making but do not meet inventory evaluation requirements. Until something better comes along, the unit standard cost will be used for inventory evaluation.

Customer Service

If the item is to be sold in the marketplace, the desired service level must be determined. If the item is to be used in manufacturing, the next inventory level such as an assembly is the "customer" and again the desired service level must be determined. A poor service level in the marketplace will lose customers, whereas a poor service level in manufacturing will cause inefficiencies such as shortages or late deliveries. Service levels are adversely affected by forecast error, purchased or manufactured lead-time error, and quality problems. If these errors cannot be corrected, the service level must be protected through safety stock or safety lead time. If the problem is forecast error, safety stock should be used, but when the problem is lead-time error, safety lead time is more practical. Both increase inventory levels.

Lot Size

The order quantity purchased or manufactured will affect the inventory and is dependent on the technique used for lot size determination.

The total inventory investment is the sum of each item's quantity multiplied by that item's standard cost. The expected level of each item is calculated by considering the planned safety stock and the usage rate of the lot size. The average expected inventory of an end item sold at a uniform rate will be the safety stock plus one-half of the lot size. The expected inventory of a manufactured item that is consumed immediately (a Just-In-Time concept) and has no planned safety stock will be close to zero.

The basic factors controlling the lot size are two opposite considerations. The factor calling for small lot sizes is the cost of carrying the inventory. Some of these costs include the following:

1. The money invested for inventory
2. The space taken for stock
3. The management of the inventory; material handling, cycle counting, and so on
4. The risk of obsolescence

An additional cost of large manufacturing lot sizes not considered in the past is the loss of flexibility and level product flow in production operations. Lot-sizing techniques have not taken this cost into account.

Whereas inventory-carrying costs call for small lot sizes, the factor that calls for large lot sizes is the setup cost either in the plant or at the supplier's facility. Most lost-sizing techniques attempt to balance the effects of carrying costs and setup costs. The advent of Just-in-Time concepts has added emphasis to reducing setup time and, therefore, reducing lot sizes.

The frequency of ordering is a function of the usage rate and the lot size. Some lot-sizing techniques, such as fixed period, will vary the quantity while maintaining a constant frequency of reordering. The standard economic order quantity (EOQ) formula calls for a fixed quantity while varying the frequency. Other more sophisticated techniques vary both the quantity and frequency of ordering.

The determination as to when to order an item is dependent on the following:

1. When the item is required. The time of need for an end item is that point in time when there is a planned customer requirement or an anticipated inventory reduction to a safety stock level. The time of need for an assembly component or a raw material is that point in time when there is a planned requirement for initiating an assembly or manufacturing order.
2. The purchasing or manufacturing lead time of the item. This is the time between initiating the ordering process and the receipt of the item.

THE PRODUCT

In order to determine the appropriate activities and procedures for inventory control, the nature of the items to control must be defined.

Finished Goods—A finished good is an end item sold to customers from either stock or made to order. It may be an assembly as complicated as an automobile or as simple as a screwdriver. It may also be a single unit such as a washer. The end item's manufacturing and stocking plans may be dependent on forecasts and/or customer orders. It will be sold through a distribution channel.

Manufacturing Items—These items are required for manufacturing an end item. They consist of raw material such as bar stock or castings, component parts (either purchased or manufactured), and subassemblies. These items may be in their completed state or in process.

Service or Repair Parts—These are component parts or subassemblies that are required for end item manufacture but are also sold individually through a distribution channel.

PRODUCT DEMAND

The actual or anticipated sales or demand of finished goods are based on forecasts, customer orders, or a combination of both. This type of demand is considered *independent* in that it is unrelated to the demand of any other items. The demand for manufacturing items is *dependent* in that it is related to the demand of finished goods or other higher-level manufacturing items used in the production of the finished goods. The demand of manufacturing items is calculated through the bill of material structure of the end items to be produced. In the Bill of Material shown in Figure 2-1, the demand for item A is independent, whereas the demand for the other five items is dependent.

The usage of the independent demands of finished goods will often be at a steady rate, whereas the dependent demand for manufacturing items will be intermittent. This is caused by lot sizing of the items in the bill of materials structure. An example of the effect of lot sizing on dependent demand usage rates is as follows:

	Demand Rate
1. End item A sells at a rate of 10/day and is produced once a week in lots of 50.	10/day
2. Subassembly B goes into A and is assembled in lots of 100 every 2 weeks	50/week
3. Machined component C goes into B and is produced with an EOQ of 800 every 16 weeks	100/every 2 weeks
4. Casting D is purchased for component C	800/every 16 weeks

The demand rate of the component is dependent on the lot size and production frequency of the parent.

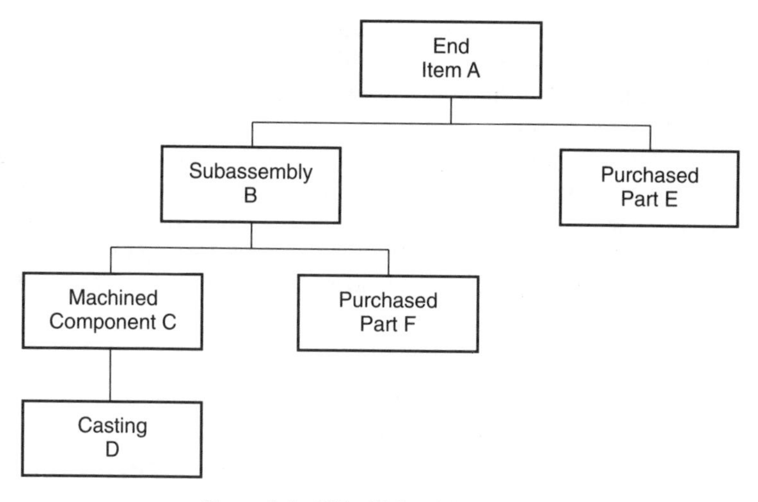

Figure 2-1. Bill of Material example.

Control Systems

Because the drivers of finished goods and manufacturing inventories differ in nature, different inventory control systems are required.

1. A *reorder point* system is used for finished goods (independent demand) that have a uniform continuous usage rate.
2. A *time-phased order point* is used for finished goods and service parts when the usage rate is not uniform. Time-phasing logic is similar to MRP and will be discussed in future chapters.
3. *MRP (Material requirements planning)* is used for manufacturing items (dependent demand)
4. *DRP (distribution requirements planning)* is used for relating branch warehouse demand to the manufacturing facility.

The demand generated through the time-phased order point system can become the driver or controller of the MRP manufacturing system. This is accomplished through the use of the master production schedule. Finished-goods inventory planning is driven by customer demand, either anticipated through a forecast or customer orders. The goal is to maintain an inventory consistent with a defined level of customer service and a realistic inventory investment.

Manufacturing inventory planning is product-driven based on calculating requirements derived from the product structure of planned finished-goods production. The service goal is to have components available for planned production of the parent item. The inventory investment goal of manufacturing items is to dispose of all the inventory to the next level. Large lot sizes and safety stocks at the component level run counter to minimizing manufacturing inventory. The reduction of these negatives must be addressed in the execution of the manufacturing process and not in the planning system. Whereas finished-goods planning is forecast-driven and manufacturing item planning is based on dependent demand calculations, service or repair parts will be forecasted based on independent demand and then be added to the manufacturing item demand.

THE REORDER POINT

The reorder point (also referred to as the order point, statistical order point, trigger) is that predetermined inventory level at which replenishment action is called for if the on-hand plus on-order quantity drops to or below that level. The reorder point system is item based on forecasted usage assuming independent and uniform demand.

The reorder point calculation is as follows:

Reorder Point = Anticipated demand during lead time + Safety stock

The lead time is the sum of the following:

1. Supplier or manufacturing lead time
2. The review period such as daily, weekly, or monthly
3. Purchase order or shop-order preparation time
4. Receiving and inspection lead time

The safety stock may be based on sophisticated statistical formulas such as measuring the standard deviation of a second-order exponential smoothing forecast and relating the deviation to the lot size, lead time, and the desired service level. On the other hand, the safety stock quantity may be a fixed number that represents an expected demand over a predetermined period of time, such as 2 weeks.

Plotting the inventory quantity versus time with a reorder system assumes the conventional sawtooth profile.

Figure 2-2 shows product A and is based on

Forecasted usage rate = 100 units/week
Lead time = 4 weeks
Lot size = 700 units
Safety stock = 200 units
Reorder point = 100 (4) + 200 = 600 units

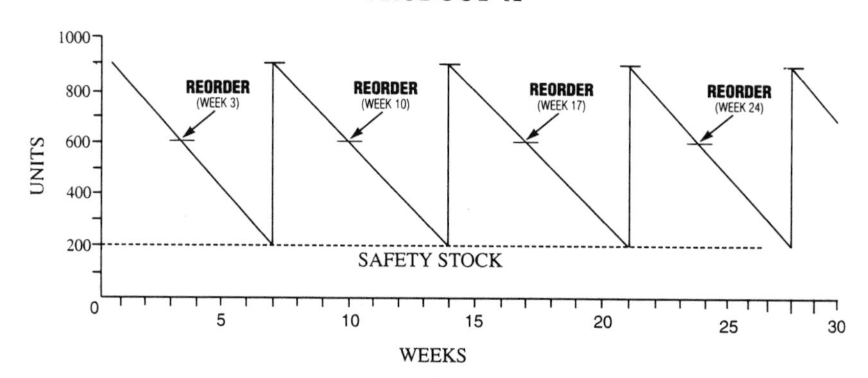

Figure 2-2. Reorder system "sawtooth" curve.

The anticipated stock levels are

Minimum stock = Safety stock = 200 units

Maximum stock = Safety stock + Lot size = 900 units

Average stock = Safety stock + ½ (lot size) = 550 units

Although the lot size does affect the quantity carried in inventory, it does not enter into the reorder point calculation. In the above example, if the lot size was 5200 (1 year's worth), the reorder point would still be 600 units.

The reorder point is relatively easy to understand, calculate, and manage. If all usage was independent and used in a consistent uniform rate, the reorder point technique would work well in inventory management. In the real world however, the demand rates do not react in such a consistent manner. For example, assume that product A above is machined from B. The supplier of B has established a lot size of 1200 and has quoted the best price at that quantity and with a lead time of 2 weeks. If the manufacturer of product A is on a reorder point system and experience has shown the 2-week lead time to be reliable, the controls for B would be

Lot size = 1200

Safety stock = 0

Forecasted usage rate = 100/week

Lead time = 2 weeks

Reorder point = 2(100) = 200

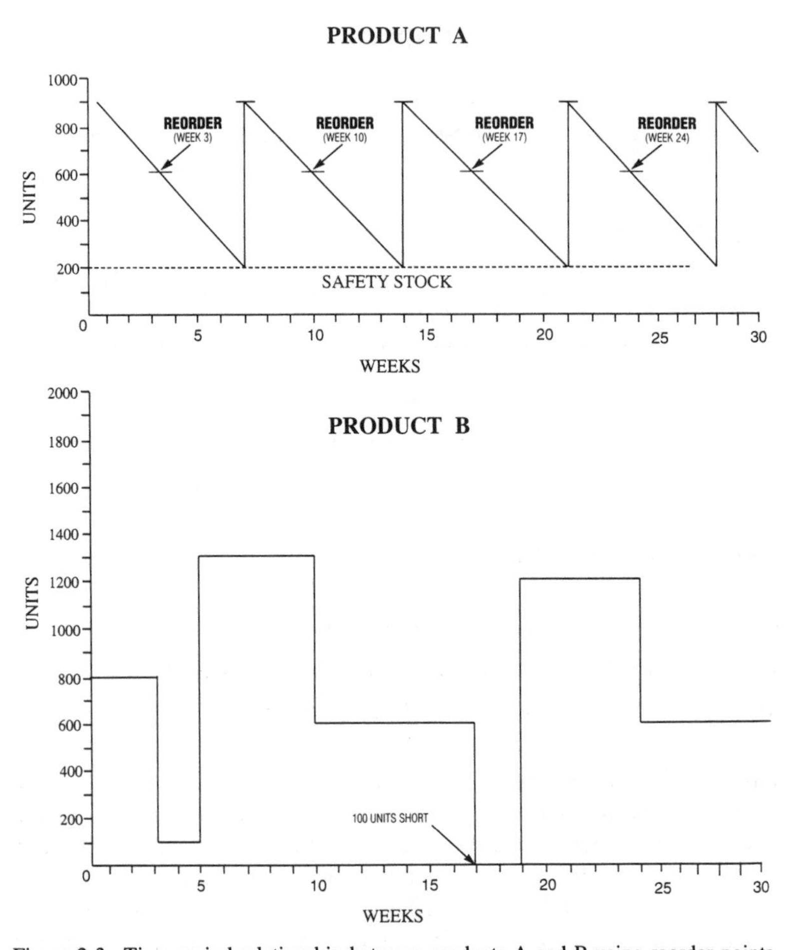

Figure 2-3. Time period relationship between products A and B using reorder points.

Relating B to A when there are 800 units of B in stock at week 0 is shown in Figure 2-3.

Two problems are shown:

1. In week 5 and 19, the replenishment order for B arrives 5 weeks before it is planned to be used—a timing problem. This has created a costly excessive stock condition.
2. In week 17, a production order to produce 700 units of product A calls for 700 units of B when the stock level is 600. The reorder point of the component is less than the lot size of the parent.

The conditions of excessive stock on one hand and the probability of shortages on the other are both addressed by MRP (material requirements planning). The quantity and timing of component requirements are based on the needs of the parent.

MATERIAL REQUIREMENTS PLANNING

Material requirements planning (MRP) is a set of techniques that calculates the requirements for all items structured in a bill of material. The calculated requirements are based on the quantity and timing requirements of the end items listed in the master production schedule. The item requirements calculation is based on the master production schedule (MPS), the bill of material file, and the item master file.

The master production schedule states the needs of the end items (an anticipated build schedule) or the timing and quantity needs of modules or artificial grouping of items. The master production schedule will consider forecasts, orders, inventories, safety stocks, and capacities. The bill of material defines by end item all related subassemblies, parts, and raw materials within the structure. This definition is level by level and defines the parent–component relationship. The end item bill may be restructed for planning purposes. Restructuring procedures and options are explained in Chapter 3. The item master or record will contain planning data of each item such as the lot-sizing rule, lead times, and inventory status.

The material requirements program will translate the master production schedule into time-phased net requirements and planned coverage for each item. The net requirements for each time period is based on the gross requirements, inventory on hand, and inventory on order. The planned schedule is based on net requirements, lot size, and lead time. If MRP were used for calculating the requirements of product B, rather than the reorder point, the planning profile would be as shown in Figure 2-4. Notice that the plan for product A is unchanged.

Comparing the results of MRP with reorder point logic for the requirements of the dependent component—product B—shows the following:

1. With MRP, the replenishment order goes to stock in week 10 when it is needed, rather than arriving 5 weeks early.
2. In week 17, when the requirement is 700 and the available stock 600, a replenishment order is called for rather than having a 100-unit shortage as would have occurred with the reorder point.

Although this example shows the accurate timing and quantity planning possible with MRP, it also shows the negative effects of component lot sizes larger than the parent's.

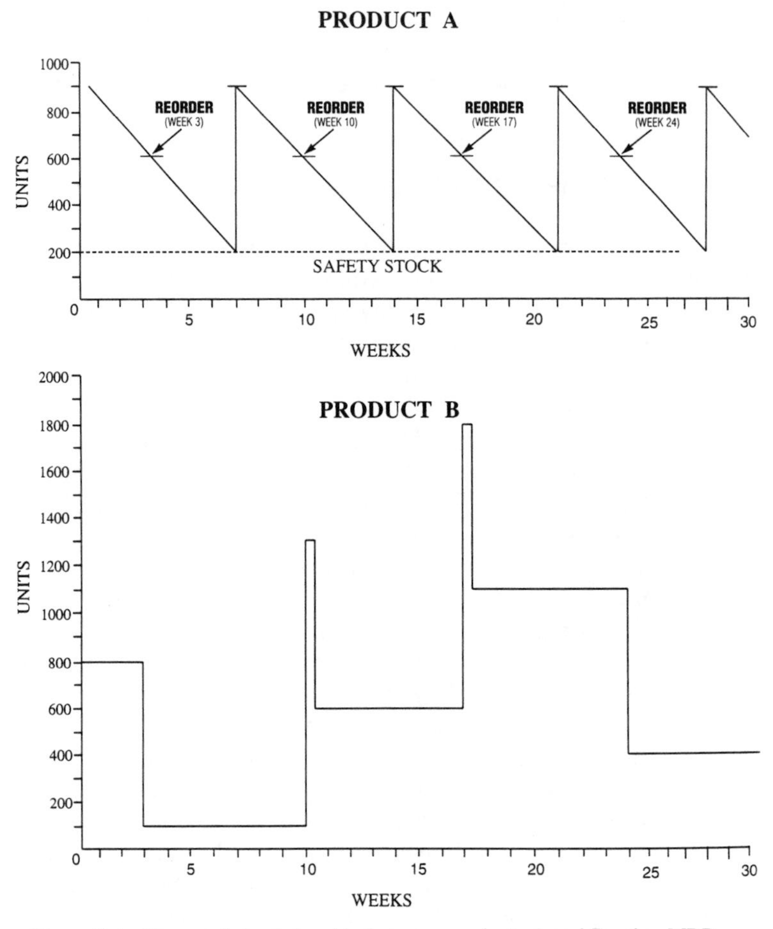

Figure 2-4. Time period relationship between products A and B using MRP.

The key word in MRP is "planning." The output of the system are recommendations to

1. Release orders
2. Reschedule orders
3. Cancel orders

Planning activities, either purchased or manufacturing, are only half the task. The other half is "execution." MRP plans must not only be accurate but also realistic. The accuracy of material requirements planning is based on understanding the principles and logic of the system, as well as the effectiveness of the data-base files.

To make the MRP system work, the plans must be realistic. Realism is attained through the principles of MRP II (manufacturing resource planning). MRP II considers the output of MRP and relates the requirements to capacity, available material, and time. When all manufacturing functions are considered and adjustments made, "closed-loop" MRP is complete. The adjustments can be made to the constraints either by increasing capacity or by adjusting the master schedule which drives the MRP. The former solution is most desirable in that it will best meet customer requirements. However, if the constraints cannot be removed, the integrity of planning must be maintained through master schedule adjustments. Not doing this will add to the problems on the shop floor and further increase customer service problems.

The focus of the balance of this book will be on both MRP planning and execution.

CASE STUDY

Problems

1. A purchased product is forecasted to have a uniform sales rate of 150 units per week. The supplier lead time including review period, purchase order preparation, and receiving is 6 weeks. In order to allow for forecast variation, a safety stock of 300 units is planned. The purchased lot size quantity is 1000 units. Calculate the reorder point and the average expected inventory.
2. The lot size of the above product is reduced from 1000 units to 600 units. Recalculate the reorder point and the average expected inventory.

Solutions

1. The reorder point = anticipated demand during lead time (150×6) + safety stock (300) = 1200 units.
   The average expected inventory = safety stock (300) + ½ lot size ($\frac{1}{2} \times 1000$) = 800 units.
2. The reorder point remains at 1200 units. The lot size is independent of the reorder points.
   The average expected inventory will now = 300 + ½ (600) = 600 units.

QUIZ

1. A forecasting system is required for which of the following elements of total demand of an item?

I. Independent demand
II. Dependent demand
III. Replacement-part demand

a. II
b. I and III
c. II and III
d. I, II, and III

2. Ideally all manufacturing inventory would be
   I. considered finished goods
   II. consumed upon completion
   III. forecasted

   a. I
   b. II
   c. I and II
   d. All of the above

3. If a forecast is 50/week, the lead time is 8 weeks, the safety stock is 4 weeks, and the lot size is 800, the reorder point is
   a. 400
   b. 600
   c. 800
   d. 1000

4. The forecast is taken into account in
   I. distribution inventory
   II. MRP
   III. the master schedule

   a. I
   b. I and II
   c. I and III
   d. I, II, and III

5. In bicycle manufacturing and distribution, the demand for a tire sold for replacement as well as used in assembly would be considered
   a. independent
   b. dependent
   c. mixed
   d. none of the above

6. The master schedule takes into account
   I. the forecast
   II. finished goods inventory
   III. order backlog

   a. I
   b. I and II
   c. I and III
   d. I, II, and III

7. Distribution requirements planning (DRP) is used for
   a. manufacturing items with dependent demand
   b. purchased components
   c. relating branch warehouse demand to the manufacturing facility

8. If 80 are required and there are 50 in stock, 30 is the
   a. gross requirement
   b. available inventory
   c. net requirement

9. A time-phased order point will be used for
   I. finished goods
   II. a uniform continuous usage rate
   III. a nonuniform usage rate
   IV. dependent demand items

   a. I and II
   b. I and III
   c. I and IV
   d. IV

10. Safety stock may be based on
   I. a predetermined quantity
   II. anticipated usage for a given time period

   a. I
   b. II
   c. Either I or II
   d. Neither I nor II

BIBLIOGRAPHY

APICS Dictionary, 7th ed., Falls Church, VA: American Production and Inventory Control Society, 1992.

Fogarty, D. W., Blackstone, J. H., Jr., and Hoffman, T. R., *Production and Inventory Management*. Cincinnati: South-Western Publishing, 1991.

Orlicky, J., *Material Requirements Planning*. New York: McGraw-Hill, 1975.

Plossl, G. W. and Wight, O. W., *Production and Inventory Control*. Englewood Cliffs, NJ: Prentice-Hall, 1967.

Sweeny, A., *Accounting Fundamentals For Non-Financial Executives*. New York: American Management Association, 1972.

Vollman, J. E., Berry, N. L., and Wybark, D. C., *Manufacturing Planning and Control Systems,* 3rd ed. Homewood, IL: Richard D. Irwin, 1992.

3
Product Definition

In order to plan production, the finished product(s) must be defined in detail. The bill of material is a list of all items that make an assembly and is structured level by level so as to show the parent–component relationship. Each item in the bill of material must be specifically identified. The parent–component relationship must be stated in a manner that will satisfy purchasing, fabricating, and assembly requirements. The details of how each item is manufactured is found in the process routing file. The data required for the routing file is dependent on the nature of the manufacturing process. This information is most critical in that it is the link between planning and execution.

ITEM IDENTIFICATION

Every identifiable unique item that is part of the manufacturing process must carry a unique part number. If a casting is purchased, machined, and then painted one of four colors, six part numbers would be required, as shown in Figure 3-1.

Because the shape and cost of the four painted castings are the same, it might be thought that one finished painted casting part number would suffice. This would not work, as the inventory system must distinguish each unique item. In reviewing Figure 3-1, part No. 2 for the machined but unpainted casting is required if the castings are machined in a single lot, stored in inventory (even for a short duration), and then taken from inventory and painted. If the castings are machined and painted as part of a two-operation process, the part number for the unpainted machined casting would not be required. Figure 3-2 shows the parent–component relationship in this situation.

The configuration or bill of material structure, as shown in Figure 3-2, has the advantage of one less part number to be maintained and one less level in the bill of material. However, the manufacturing process consistent with this simplified structure requires that the castings be machined and painted as part of a single routing. A machined but unpainted casting without

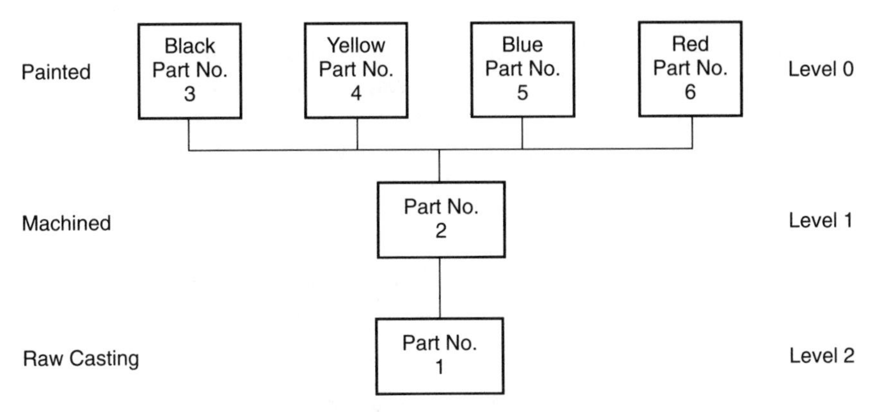

Figure 3-1. Three-level item identification.

a unique part number cannot be carried in inventory. The elimination of part No. 2 and reducing the number of levels in a structure is called "collapsing" the bill of material. It is an example of how the method of processing can affect the bill of material structure.

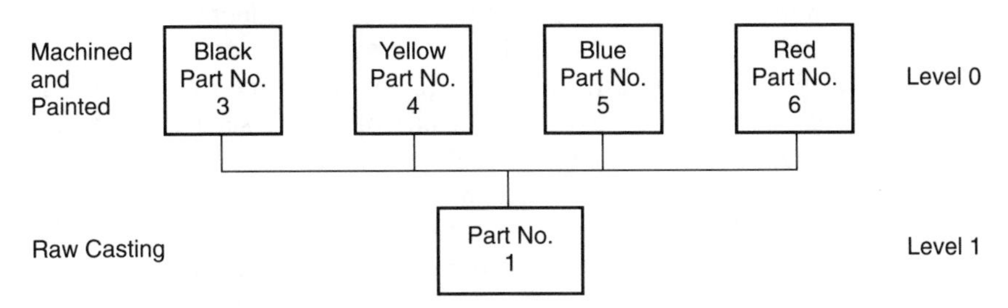

Figure 3-2. Two-level item identification.

Part Number Significance

The part numbering system used to identify unique items can be significant, nonsignificant, or semisignificant.

A significant part numbering system assigns a part number that will be used not only to identify and control the item but also to define or partially define the item. An example would be a series of alphabetic or numeric numbers assigned to bolts used in engine block assemblies. A second example would be a series listing aluminum castings. The advantage of a significant part numbering system is that users are often more comfortable with the descriptive feature of the part number, as it allows recognition of the item.

The disadvantages are that a greater number of digits (a larger part number) are required, the allocated grouping of unused numbers may be used up after a period of time, and there can be confusion in defining the grouping. An example would be the engine bolt that can be used in other assemblies. The addition of an entire new product line would also create problems.

A nonsignificant system applies numbers in sequence that are used to identify an item but in no way describes the item. The advantages of a nonsignificant system is less digits (a shorter part number) are required and there are no concerns about number allocation and descriptive categories. The disadvantage is the user's initial discomfort with a nondescriptive number and the increased possibility of more than one number being assigned to the same item.

Some firms have gone to a semisignificant system in an attempt to gain the advantages of a significant system but to avoid the problems. An example is the use of random or nonsignificant part numbers with an "A" prefix for assemblies. Another example would be that all items in a numerical series such as 70000 would be class C hardware.

Part Number Classification

Not to be confused with part numbering systems, which identify each item used in an operation, are part classification systems. A part classification system is a subsystem useful for both design and manufacturing engineers. Part classification software defines all details of a part such as material, shape, and required processing. The level of detail for complete definition can require up to 28 digits. Classification can assist the design engineer in reviewing existing parts and consequently save design time and reduce the risk of part proliferation. The lack of parts classification coupled with the rapid output rate of CAD-CAM (computer-aided design/computer-aided manufacturing) drawings, if not closely monitored, can cause problems of redundant parts proliferation.

Once the item is designed, a parts classification system can assist the manufacturing engineer in creating a process routing consistent with similar parts in the manufacturing operation. Parts classification is also useful in process review in order to plan manufacturing cells or clusters.

THE BILL OF MATERIAL

A single-level bill of material will list the parent, the component(s), and quantities directly used to make the parent. Therefore, a single-level bill will consist of two item levels. Examples of a single-level bill of material is a clock assembly (the parent) consisting of a clock, a base, and two fastener screws. Another example of a single-level bill of material is a machined part (the parent) and the purchased casting required for machining. A multilevel bill of material will list all components and their quantities required for an assembly and is structured level by level taking into account all purchased, fabricated, and assembled part numbers. This structure indicates the all-important dependent relationships of the components to the parents. A part such as a machined casting or a subassembly will be both a parent to its components and a component to its parent at the next level in the structure. The bill of material, the list of items in an assembly, can be in summarized form listing all parts and required quantities but not identifying where-used (parent–component) relationships, such as what parts are used in what subassemblies.

The bill of material is normally a document written and maintained by design engineering. The structure of the bill is based on the perspective of the design engineer and is after the fact in that it is defining a product that has been designed. It can, therefore, be assumed that the bill of material is not needed to design a product, but it is required to manufacture the product. Based on the needs of the manufacturing operation, it may require restructuring the bill while not changing the product itself.

Requirements of the Bill

The bill should allow the following:

1. *Ease of Product Forecasting*. If a product consists of four major modules and there are three versions of each module, the combination of the various modules would allow for 81 final products (3×3×3×3). Forecasting at the modular level would call for 12 items (3+3+3+3) to be forecasted rather than 81. The less items forecasted the higher the degree of accuracy. In this example, if the end product was not structured in modular form, restructuring would be needed.
2. *Ease of Master Scheduling*. As with forecasting, the less the number of items to control, the more workable the plan. If an option is called for on 40% of a number of end items, it is practical to plan the option in a stand-alone manner rather than to structure it to end items and estimate its usage rate for each end item.
3. *Consistency with the Manufacturing Process*. If a subassembly is really just a separate step in the assembly operation, that subassembly should not be listed as a separate level but should be reviewed for possible elimination from the structure.
4. *The basis for Product Costing*. Some items such as packaging may not be product controlled, but they should still be listed on the bill for costing purposes.

5. *Order Entry and Final Assembly Scheduling*. Even though forecasting and master scheduling may be done at the modular level, the customer may order by model number and the final assembly schedule must be specific end items.

The above requirements may be in conflict with each other, and for that reason, separately structured bills may be used in the planning and manufacturing of an item.

Multilevel Product Bill

The number of levels in a bill of material will be a function of the complexity of the product. An item may be in the structure at more than one level. A simple example of a product bill is a table lamp as shown in Figure 3-3.

As the lamp is structured there are 15 part numbers required, 3 of which are subassemblies. There are four levels in the bill, the wire and plug being at the lowest level and going into the wire assembly which, in turn, goes into the socket assembly. If the process could be changed so that the lamp would be assembled in a single operation such as an assembly line, the base, socket, and wire assemblies could be eliminated and the bill collapsed from four to two levels. The end result would be a single-level bill as shown in Figure 3-4.

Phantom Bill of Material

If the manufacturing process allows elimination of subassemblies such as the base, socket, and wire subassemblies in the lamp assembly shown in Figure 3-3 but there is a possible need to stock the subassemblies due to service usage or occasional overruns, the part number for the subassemblies cannot be eliminated. The subassemblies remain in the bill structure but are coded as phantom bills. Other names for this technique are pseudo or transient bills of material. While the phantom bill remains in structure, the MRP software blows through this level as though the phantom subassembly does not exist. The lead time is specified as zero and the lot sizing is lot for lot. If there is stock of the phantom subassembly, the system will consider its availability and call for a reduced number of components.

PLANNING BILLS OF MATERIAL

The nature of the product will dictate the bill of material structure best suited for planning purposes. If the end item is relatively simple without a number

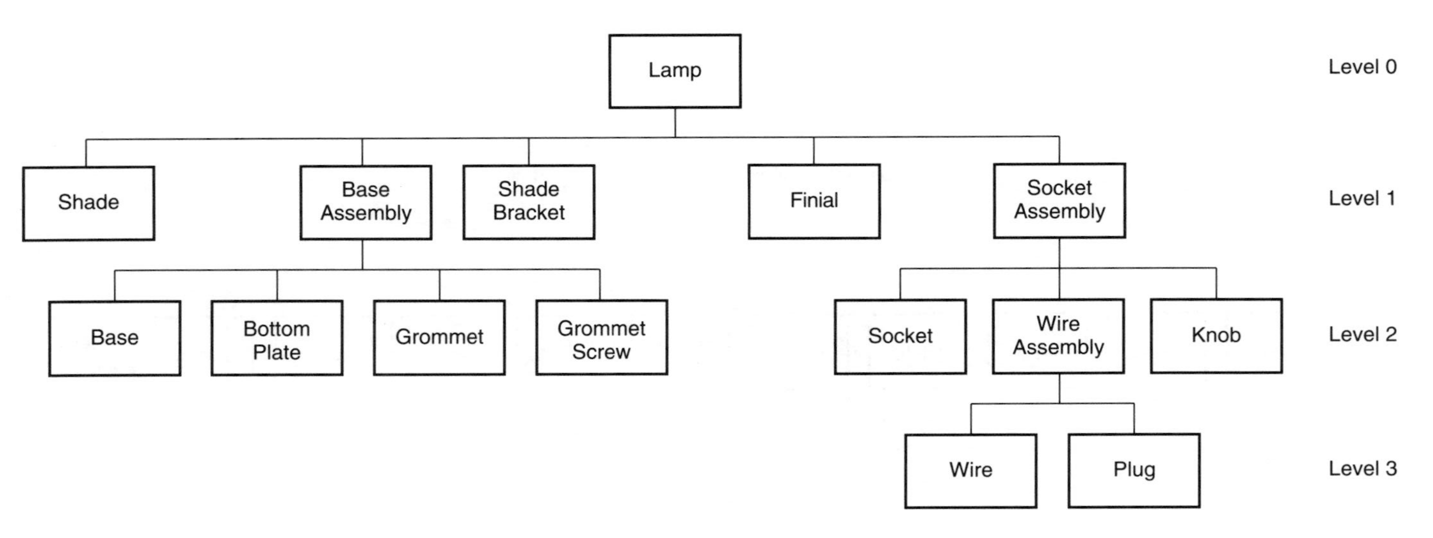

Figure 3-3. Table lamp, multilevel product bill.

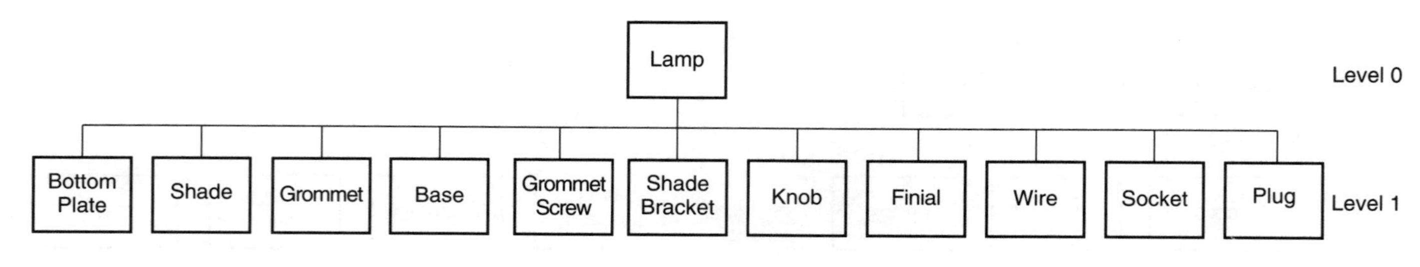

Figure 3-4. table lamp, single-level product bill.

of options to be considered, a product bill of material similar to the lamp shown in Figure 3-3 or 3-4 would be most practical because the structure meets the five requirements listed earlier in this chapter. If the end product bill of material will not meet the requirements, restructuring to a planning bill format is needed. A planning bill is defined as an artificial grouping of items used for the purposes of forecasting and/or master scheduling.

A modular bill is a type of planning bill in that it will be manufactured as defined in the bill. It might be considered a major subassembly. Examples of modules would be automobile engines or transmissions. A second type of planning bill is a parts listing of all common components for a product or product family. This is known as a common parts bill. A super bill is a type of planning bill that ties together modular and/or common parts bills to define a product or product family at the top level.

An example of planning bill restructuring using modular and common parts bills is shown by a review of the product bill of material of the lamp shown in Figure 3-4. Item variations might be as follows:

Shade = 3—12, 14, 16 inches

Base = 3—gold plated, silver plated, glass

Bottom plate = 1

Grommet = 1

Grommet screw = 1

Shade bracket = 1

Finial = 1

Socket = 2—one Way and three Way

Wire = 1

Plug = 1

Knob = 1

The possible combination of lamps is 18 ($3 \times 3 \times 1 \times 1 \times 1 \times 1 \times 1 \times 2 \times 1 \times 1 \times 1$) and, therefore, 18 product bills of material.

A modular bill restructuring would be as shown in Figure 3-5. Rather than forecasting and planning 18 lamps, the modular structure calls for 8 items of control. Additional restructuring would allow common items structured to a module such as the grommet or plug to be structured with the common parts grouping (or kit). The common parts forecast is the sum of the total number of lamps planned through the module forecast. Overplanning the common parts option is a method of safety stock allowance.

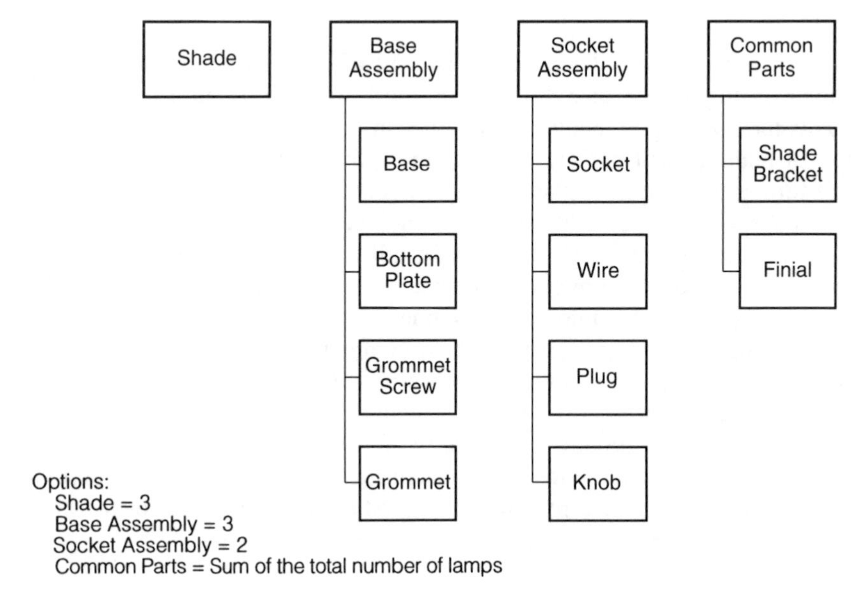

Figure 3-5. Table lamp modules.

Planning Bill Relationships

Although planning bills simplify forecasting and master scheduling, an additional advantage is that planned components are not committed to exact products until the final assembly of the product is scheduled. There will be a final product bill structure that may differ from the planning bill structure, but the two bills must be consistent and relate to each other. An example would be a planning system based on the four lamp modules shown in Figure 3-5 but with a final assembly product bill as shown in Figure 3-6. This structure assumes a two-step assembly operation—mechanical and electrical.

THE PROCESS ROUTING

The necessary details of how to manufacture an item are found in the routing file. Basic information are the operations to be performed, in what sequence, and at what work centers. Standards for setup and run times as well as tooling information are also included.

Routing data are used in capacity planning and shop-floor systems. As with bills of material, there have been efforts to simplify the process routings, but again the routing is merely a recording of the required process. The best way to collapse the routing is to standardize and simplify the process.

Figure 3-7 is an example of a process routing for the fabrication of a part

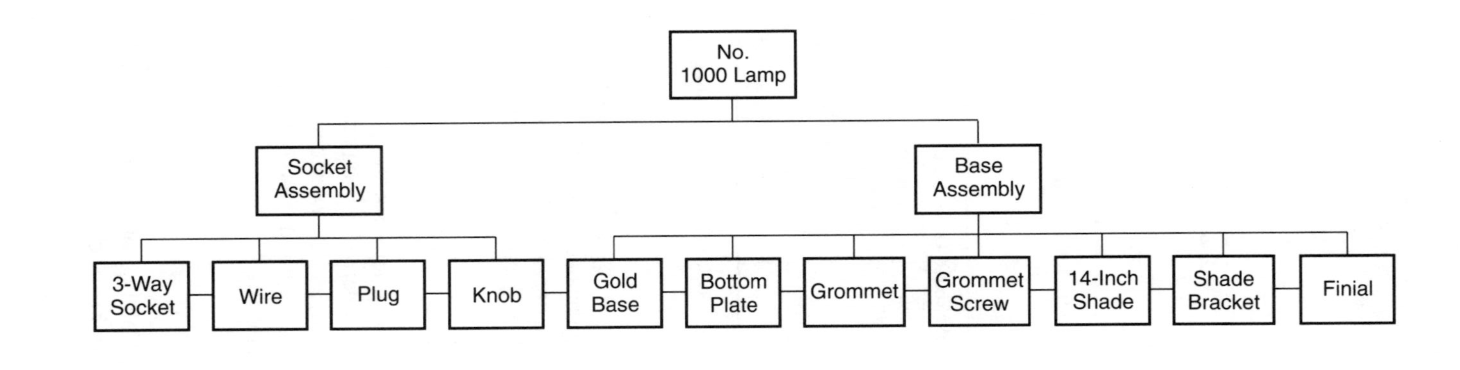

Figure 3-6. Table lamp, model 1000 assembly bill.

Part No. XXXX Guide Block				
Operation	Work Dept.	Work Center	Run Time hrs/100	Setup Hours
10	Saw	102	.8	.6
20	Mill	204	2.4	.4
30	Mill	205	3.1	.2
40	Drill	301	1.1	.5
50	Insp.	602	.4	.1

Figure 3-7. Job-shop routing file.

Part No. XXXX Guide Block			
Work Cell	Cell No.	Operators Assigned	Output Rate Pieces per hour
Flat Block Fabrication	482	1	13
		2	26
		3	32

Figure 3-8. Dedicated work cell routing.

produced in a job shop organized by a functional or departmental layout. If the part were produced in a dedicated work cell, the routing would be much less detailed, as shown in Figure 3-8. The more detail in a routing is an indication of the greater amount of scheduling control required for production.

Work Center Data

The work center master file contains specific data necessary for capacity determination, such as the number of scheduled shifts, the number of machines, the work days per period, and the work center utilization and efficiency factors. Also listed will be the required data to determine queue time allowances for scheduling calculations.

File Organization

The bill of material file (the defining of the product) and the routing file (the defining of the process) have historically been two distinct data files. This was due to the following:

1. Design engineering controlled the bill of material while manufacturing engineering controlled the routing.
2. Because a component may be used on many products, including the same routing data in each bill of material would use a high amount of storage space.

Changing Times

In the last 10 years, organizational concepts and philosophies within manufacturing environments have been rapidly changing. The small business unit concept is bringing functional operations together not only from an organizational viewpoint but people are now working side by side. The distinction between design engineer and manufacturing engineer is not as clear. With the increase usage of manufacturing cells rather than the departmental layout, less detailed operational information is required in the routing file. In addition, much fixed data such as tooling information is being taken off the file and moved to the shop floor.

With less detail, more teamwork, and increased computer storage capacity, it has been suggested that the routing file be eliminated and the routing data become part of the bill of material. This concept will be discussed in greater detail later in this book when planning and execution details are reviewed. The software ramifications of this change will require much study but should be undertaken.

CASE STUDY

Problems

A desk manufacturer's product line consists of 32 models. There are four different size tops, two different size frames, and four combination drawer arrangements ranging from two to five drawers. With a relatively low total demand, the forecasting of 32 models has been difficult and inaccurate. A decision is made to plan at the modular level.

1. Will bill restructuring be required?
2. Forecasting of 32 end products will be reduced to how many items?
3. At what level of the MRP II system will the modules be utilized?
4. Will there be any use for the 32 end item bills?

Solutions

1. Bill restructuring will not be required as the modules are already unique. Items within the end item bill—one level down.

2. Forecasting at the modular level will be 4 (tops) + 2 (frames) + 4 (drawer combinations) = 10 modules.
3. The forecasted modules will be planned in the master schedule.
4. The end item bill will continue to be used for order entry, final assembly, and product costing.

QUIZ

1. The bill of material is primarily a document for design engineering.
   a. True
   b. False

2. The bill of material is used for
   I. MRP
   II. product costing
   III. defining the product

   a. I and II
   b. I and III
   c. II and III
   d. All of the above

3. A separate part number must be generated when
   I. the item has changed identities
   II. the item is a parent
   III. the item is a component of more than one parent

   a. I
   b. I and II
   c. I and III
   d. All of the above

4. A transient subassembly or phantom will be
   1. coded as such
   II. have a lead time of "0"
   III. have a "lot-for-lot" lot size

   a. I
   b. I and II
   c. I and III
   d. All of the above

5. If a subassembly is immediately consumed in the assembly operation but is carried in stock for service usage, that subassembly can be
   a. modularized
   b. coded as a phantom
   c. eliminated from the bill
   d. purchased

6. A finished item such as a bolt which is manufactured directly from purchased steel rod would be defined by a
   a. phantom bill of material
   b. planning bill of material
   c. single-level bill of material
   d. multilevel product bill

7. The details of how to manufacture an item are found in
   a. the item master file
   b. the bill of material file
   c. the planning bills
   d. the routing file

8. The bill of material can be restructured for purposes of forecasting.
   a. True
   b. False

9. Planning bills assist in
   I. ease of forecasting
   II. master scheduling
   III. costing out finished goods

   a. I and II
   b. I and III
   c. II and III
   d. All of the above

10. The bill of material is used for
   I. MRP
   II. definining the process
   III. cycle counting

   a. I only
   b. I and III
   c. II and III
   d. All of the above

BIBLIOGRAPHY

APICS Dictionary, 7th ed., Falls Church, VA: American Production and Inventory Control Society, 1992.

Fogarty, D. W., Blackstone, J. H., Jr., and Hoffman, T. R., *Production and Inventory Management*. Cincinnati: South-Western Publishing, 1991.

Goldratt, E. M., *The Haystack Syndrome*. Croton-On-Hudson, NY: North River Press, 1990.

Orlicky, Jr., *Material Requirements Planning*. New York: McGraw-Hill, 1975.

Lunn, T. with Neff, S. A.; *MRP—Integrating Material Requirements Planning and Modern Business*. Homewood, IL, Business One Irwin, 1992.

Stoddard, W. G. and Rhea, N. W., We Need to Change the Role of MRP in Manufacturing. *Target,* Vol. II(1), Spring 1986. Wheeling, IL: Association for Manufacturing Excellence.

4
The Master Production Schedule

The planning process for MRP II starts with a business plan which is a long-range strategy plan based on forecasted sales income and broad-based product family groups. The business plan anticipates future production facilities and personnel requirements as well as financial considerations. The production plan developed from the business plan is of an intermediate time period and, based on family groups, will be used for the initial test of available capacity (rough-cut capacity planning). When determined to be realistic relative to capacity, the production plan is converted to specific products to be utilized in the master production schedule.

The master production schedule is a detailed plan of production. If the product is an uncomplicated assembly such as a stapler, the master schedule would be expressed as an anticipated assembly schedule over a specific planning horizon such as 1 year. If the product is a single processed item such as a washer, the master production schedule would be expressed as the anticipated production schedule over the planning horizon. A complicated assembly such as an automobile will be expressed in a master schedule representing the anticipated build requirements of modules such as engines and transmissions as well as options such as air conditioners and CD players.

The master schedule is not a sales forecast but an adjusted forecast of production demand. The production demand considers the sales forecast, existing customer orders, existing inventory, desired safety stocks, and stocking levels. Once the production demand is determined, it may be adjusted due to lot sizes, capacity considerations, and load leveling. The master production schedule is not a wish list but must be compatible with available material, realistic lead times, and capacity.

The master production schedule (MPS) drives the MRP system by referencing the bill of material and inventory files to determine the material requirements of all components. Over an extended horizon, the MPS is the basis of estimating future resource requirements such as labor, machine tools, space, and cash.

THE FORECAST

Manufacturing planning is normally based on a forecast of future demand. The word forecast is derived from "reaching back and throwing forward." In other words, predicting the future using information from the past. Lacking useful information from the past, the forecaster may be forced to make a judgmental prediction based on extrinsic knowledge. An example of a judgmental prediction would be that of a forecast for an environmental cleansing product based on the anticipated passage of stricter air pollution laws. Available leading indicators are extrinsic or outside influences that can be used in forecasting such as using housing starts to forecast sales of household fixtures.

Statistical forecasting assumes that the future will be like the past. Forecasting future replacement tire sales may be based on intrinsic data such as past tire sales or extrinsic data such as past gasoline sales. The replacement tire forecast may also be adjusted due to extrinsic knowledge, such as improved quality allowing for longer tire life.

Long-term forecasting is used for the business plan, whereas intermediate and short-term forecasting will be used in determining the production plan and the master production schedule. Intermediate forecasting covering 6 months to 1 year is required for planning cash flow, personnel, subcontracting, and supplier communications. The short-term forecast details are required for planning material requirements, order priorities, and short-term capacity requirements.

The Principles

There are certain rules to keep in mind relative to forecasting:

1. The forecast will be more accurate for groups. Total units or dollars sold are easier to forecast than to forecast by specific products. The forecast that a company will sell a total of 2500 lathes next year might be useful for cash flow information, but there must be more product details for the MPS. If there are 100 different lathes produced from combinations of 12 modules, forecasting the 12 modules will be more accurate than forecasting 100 lathes and will meet the needs of the MPS.
2. The forecast will be more accurate for the short term. "The farther out you go, the wronger you are." With this fact in mind, every effort should be made to reduce the cumulative lead time required for production of the products. The cumulative lead time is the total planned length of time to produce an item. It is the longest combination of events, the critical path, necessary for completion.

3. The forecast will be wrong. Although there will be forecast error, it is most important to have an estimate of that error. Through mathematical techniques, it is possible to estimate the probability of error. For example, the forecast of weekly demand may be 100, but based on past deviations from average, the actual demand may be expected to vary by plus or minus 6, 98% of the time.
4. Test the forecasting method before using. There are many models to use for forecasting and it is recommended to test the various techniques based on the same past history. The technique or model which worked best in the past will most likely work best for the future.
5. The forecast is no substitute for actual demand. As there will be a degree of error in any forecast, reducing lead time as much as possible, so as to allow actual demand to have a greater impact on the MPS, is most desirable. If the assembly schedule is based on customer orders, there should be no excess end items produced unless required for load leveling or future capacity considerations.

Forecasting Problems

The objective approach required for forecasting can be lost when there are conflicting objectives involved and not understood. The forecast should not be confused with a sales quota or goals, but rather what is the most reasonable expectation of activities. Gathering information from the field via the sales department has the advantage of hands-on experience, but this information must be analyzed and perhaps adjusted due to the natural optimism of many sales people.

Care must be taken to forecast the proper items that will interface with those items to be used in the master schedule. If an automotive MPS is based on modules and options such as engine, transmissions, and air conditioners, the forecast must be for those items and not just how many cars will be sold.

The forecaster must be aware of extrinsic information that will affect the forecast. Going back to the replacement tire example, a sound statistical forecast based on past history will be of little use if the quality of tires now in use will allow longer tire life. The forecaster must be aware of this fact and make the proper adjustments.

Areas of Consideration

Planning the forecasting function is critical. The first determination is what to forecast. For long-range financial planning, total dollar sales will suffice, but if production capacity is in question, a forecase for product families will be required. Forecasts to be used in the determination of MPS demand must be in those units or modules utilized in the MPS. The MPS demand may not only cover end items or modules but also service or replacement parts. In this case, the forecasted requirement of service and replacement parts must be included.

There must be an understanding of what factors, both intrinsic and extrinsic, are applicable. A tire may wear out as a function of use, a gasket wear out as a function of time, and a windshield break as a random chance of a flying piece of debris.

The availability and timing of data must be understood. If late shipments are common or end-of-month pull-ups are the practice, shipment data does not reflect customer demand. In this situation, collecting demand based on the customer requested shipping date would be more accurate. The problem in this instance is that actual shipment data are more easily accessible. A second timing consideration is in short-term forecasting. The forecast of March activity will be most accurate if actual February demand is considered. The reality is that collecting and linking information from order entry or shipping systems to the forecasting system, generating, reviewing, and then publishing the forecast does take time. The goal should be to shorten the forecasting cycle as much as possible so as to include as much up-to-date information as available.

An understanding of the time series pattern of the product is required. There are five patterns or combinations that may apply.

1. Linear—The activity will follow a straight-line (linear) pattern, such as the growth of hamburger sales.
2. A Trend Pattern—A pattern that indicates a trend over and beyond the linear pattern, such as a product demand that is growing due to not only population growth but to superior quality.
3. A Cyclical Pattern—Such as a product with a life cycle of 3 years and therefore a replacement cycle of 3 years. The business cycle may also be considered.
4. Seasonal Pattern—Such as lawn mower sales in the spring and snow-blower sales in the fall.
5. Random Happenings—This is irregular and not easily understood. The broken windshield, mentioned above, is an example of factors causing random happenings.

An understanding of the pattern will assist in the determination of the most desirable forecasting technique.

Forecasting Techniques

How the historical data are used for forecasting will be dependent on the patterns of past history as well as the availability of the data.

If there are no radical changes anticipated for the future and the activity is relatively constant, a simple average may be used. If the past 6 months is determined to be the base period, the following is an example:

Month	Demand
February	36
March	38
April	39
May	38
June	42
July	40
	233÷6 = 38.8 monthly forecast

If the forecaster using the above data decided to use a moving 3-month average, the forecasts would be

Month	Demand
February	36
March	38
April	39
	113÷ 3 = 37.6 monthly forecast

Month	Demand
March	38
April	39
May	38
	115÷3 = 38.3 monthly forecast

Month	Demand
April	39
May	38
June	42
	119÷3 = 39.7 monthly forecast

Month	Demand
May	38
June	42
July	40
	120÷3 = 40.0 monthly forecast

The moving-average method using the same data as the simple average indicated a positive trend in the level of activity.

If the most recent history seems to give a more accurate view of the future, a weighted average can be used. An example using the same 6-month history is

Month	Demand	Weight Factor	
February	36	0.11	3.96
March	38	0.12	4.56
April	39	0.14	5.46
May	38	0.17	6.46
June	42	0.21	8.82
July	40	0.25	10.00
		Weighted average =	39.26

Exponential smoothing is a method of forecasting based on the weighted-average technique but requiring the maintenance of two numbers only—the last forecast and the actual demand of the last period. The formula for basic or first-order exponential smoothing is

$$\text{New forecast} = \text{Old forecast} + \alpha\,(\text{Last period demand} - \text{Old forecast})$$

The α is the weighing factor where an α of 0.1 gives little weight to last actual demand and 0.3 gives much weight to last actual demand.

An example of increasing demand and an α factor of 0.2 shows

Old forecast = 80
Demand last period = 86

$$\begin{aligned}\text{New forecast} &= 80 + 0.2\,(86 - 80)\\ &= 80 + 1.2\\ &= 81.2\end{aligned}$$

If the demand were under forecast and the α factor = 0.15, the calculation would be

$$\begin{aligned}\text{New forecast} &= 80 + 0.15\,(76 - 80)\\ &= 80 - 0.6\\ &= 79.4\end{aligned}$$

The advantage of exponential smoothing is the simplicity of calculation as well as avoiding the necessity of carrying historical demand data.

When there are trend influences to be considered, second-order smoothing, which allows for the trend effect, may be used. The details of this equation will not be covered in this book. Seasonal influences can be covered by applying an index for a specific season, such as treating a given month as 12% of the annual forecast rather than 8.3%.

Focus Forecasting

With computer technology, it is now possible to test a number of forecasting techniques or models for each item to determine which technique would have been the most accurate in forecasting the last period. With focus forecasting, the most successful technique for the past is then used for the future forecast. One product may have been best forecasted for the next 3 months by increasing the previous 3 months' movement by 10%. Another product might have been best forecasted by using exponential smoothing with a .3 α factor.

Focus forecasting started by using simple models which were easily understood, such as the next quarter will be 105% of the same quarter demand last year. The techniques have since been increased to include both exponential smoothing with varying α factors and other more complex models. The accuracy of the forecasts are measured based on the mean absolute deviation (MAD) of the actual to forecast.

Error Measurements

Initial determination of the degree of variation from the forecasted number is achieved by the recording of the detailed data that determined the average. If a forecast of 4,200 was based on the following data, the standard deviation calculation would be as follows:

Period	Actual	Forecast	Deviation (D)	D^2
1	4,000	4,200	−200	40,000
2	4,300	4,200	+100	10,000
3	4,200	4,200	—	—
4	4,400	4,200	+200	40,000
5	4,000	4,200	−200	40,000
6	4,300	4,200	+100	10,000
	25,200		800	140,000

$$\text{Average} = 25{,}200 \div 6 = 4{,}200 = \text{Forecast}$$

$$\textit{Standard deviation} = \sqrt{\frac{\Sigma D^2}{N}} = \sqrt{\frac{140{,}000}{6}} = \sqrt{23{,}333} = 153$$

Based on statistics of a normal distribution, the actual number will be within three standard deviations 99.86% of the time.

In this situation, the expected range would be

$$4200 \pm (3)(153) = 3741 \text{ to } 4659$$

and 97.72% of the time, the actual number will be within two standard deviations and the expected range would be

$$4200 \pm (2)(153) = 3894 \text{ to } 4506$$

and 84.13% of the time, the actual number will be within one standard deviation and therefore the range would be

$$4200 \pm 153 = 4047 \text{ to } 4353$$

The planned safety stock will be a function of the standard deviation and the desired service level. In the above example, if the desired service level was 98%, the calculated safety stock would be twice the standard deviation or (2)(153) = 306.

A calculation simpler than the standard deviation is the mean absolute deviation (MAD), which is the average of the absolute deviations. In the above example, the MAD calculation is

$$MAD = \frac{\Sigma D}{N} = \frac{800}{6} = 133$$

The standard deviation is approximately 1.25 times MAD. In the example, the standard deviation would approximate to

$$133 \times 1.25 = 166$$

compared to 153.

The deviation from forecast should be continually measured.

Tracking Signal

In measuring forecast errors, a tracking signal will indicate bias on either side of the forecast. The tracking signal calculation is

$$\textit{Tracking signal} = \frac{\Sigma \textit{ Net forecast errors}}{MAD}$$

If tracking the difference with the previous forecast showed the following:

Period	Actual	Forecast	Deviation
7	4300	4200	+100
8	4400	4200	+200
9	4300	4200	+100
10	4100	4200	−100
11	4300	4200	+100
12	4400	4200	+200
13	4100	4200	−100
14	4200	4200	—
15	4300	4200	+100
16	4300	4200	+100

Σ Absolute deviations = 1300
Σ Net forecast errors = 700

then MAD is 1300/10 = 130 and the tracking signal is 700/130 = 5.4.

A tracking signal greater than 4 or 5 indicates a high bias and is a signal to review the forecast. If the forecast calculation was based on a simple average, the above forecast would be recalculated to 42,700/10 = 4270.

Demand Filter

Data collected for forecasting review are monitored through a demand filter. Any reading greater than 3.2 standard deviations or 4 MAD will only happen 0.07% of the time (7 out of 10,000). The demand filter will highlight for review any reading above the established value such as 4 MAD. In the original example of a 4200 forecast and a MAD of 133, the demand filters would be

$$4200 + 4(133) = 4732$$

and

$$4200 - 4(133) = 3668$$

Any readings less than 3668 or more than 4732 should be reviewed first for data entry error; and if the reading is then determined to be valid and reasonable, the forecast should be reviewed.

THE MASTER SCHEDULE CALCULATION

Demand Sources

In the development of the master schedule, various demands must be considered. In a make-to-stock environment, the forecast of future sales over the total horizon is most critical. The forecast demand must be in time periods or buckets consistent with MPS and MRP programs. The sales forecast must include not only finished product such as assemblies but also replacement parts forecasted for service or testing.

In a make-to-order environment, production is not initiated until the order is received, but a sales forecast may be required extending to the critical path in order to have raw material or purchased parts available at the time of order receipt. In an assemble-to-order environment, the sales forecast is required in order to have components available for assembly at the time of order receipt.

The customer order is the major demand source in make-to-order manufacturing and is also the demand source in the in-close periods of assemble-to-order situations. In make-to-stock or "off-the-shelf" operations, the customer order demand may or may not be considered, depending on comparing this actual demand with the forecast. The relationship of customer orders and forecasts are controlled by predetermined time fences which will be discussed later in the chapter.

Forecasted sales demand from distribution centers must be considered and care taken to understand if this demand is part of the total forecast. If the distribution demand is forecasted by time period and reported through an organized distribution requirements system (DRP), an adjustment to the total sales forecast may be required.

Other demands to be considered are desired safety stocks, planned work-in-process levels, and allowances for product shrinkage. Again, the source data must be understood. DRP requirements transmitted to the manufacturing facility should have already allowed for safety stock and inventory adjustments.

Master Schedule Adjustments

Repeating that a master schedule is an anticipated build plan or a detailed plan of production, simply scheduling anticipated demand by time period is not realistic. Production lot sizes must be considered. The production demand may be 50 units per week, but if the smallest practical lot size is 150 units, the master schedule must reflect a requirement of 150 units with a frequency of every 3 weeks.

A most painful adjustment of the MPS demand is when the system says that there is not sufficient capacity. Only after retesting, considering outsourcing, overtime, and so forth, should there be an adjustment to the master schedule. This information must be understood and communicated to all within the organization. There is a saying that has proven to be quite accurate, "Do not lie to the master schedule." Often a master sheduler may lie to the master schedule, that is, overload in order to buy time or perhaps keep his or her job.

A less painful adjustment of demand may be for the purpose of load leveling. An overload in one period may be adjusted by shifting to an underloaded period or the total load may be adjusted for the purpose of smooth, continuous flow through the manufacturing and supply chain operations.

The demand-frequency adjustment due to lot size can be automatically handled through the MPS system logic, but load leveling and capacity adjustments require management intervention.

UTILIZING PLANNING BILLS

Manufacturing Environment

In the determination of bill of material structuring strategy for master scheduling, both the nature of the product and the manufacturing environment must be considered. An uncomplicated product with relatively few components and minimal bill levels will be maintained at only the product level. An extreme example of an uncomplicated product bill would be a washer stamped out of sheet metal. For more complicated products, the product structure to be utilized for master scheduling is dependent on the manufacturing environment. Three manufacturing environments are considered.

1. Make-to-stock and sold "off-the-shelf"
2. Assemble-to-order utilizing components both purchased and manufactured, that have been purchased and/or made-to-stock.
3. Make-to-order products that may be produced from a combination of standard items and customer designed items made-to-order

Make-to-stock products would normally have a limited number of finished-goods items. The number of components and raw materials would be dependent on the product. The above-mentioned washer would consist of one component—the sheet steel—or up to two or three components if packaging is included. A product line of VCRs might consist of six finished machines utilizing a total of 600 purchased parts and subassemblies. The assemble-to-order product line will have a large number of finished goods that are assembled from a much lesser number of components and subassemblies. In this environment, the major subassemblies are called modules, that is, modular bills of material. Although the modules are fewer in number, they consist of a much larger number of components. An example of assembling-to-order might be an automotive manufacturer that assembles to dealer orders from modular planned engines, transmissions, and so on. The make-to-order product may have a simple bill (again the washer) or an expanded finished-goods line coming from a lesser number of components and raw materials. An example of the latler would be custom-made furniture.

MPS Level

In the make-to-stock environment, the master production schedule is based on the finished product and, therefore, utilizes the product bill of material. The final assembly or finished product schedule will be at the same level and be based on the same product (or end item). Bill of material master scheduling in an assemble-to-order environment is at the modular or subassembly level. The product structure utilized in this situation calls for planning bills such as modular and common parts bills. Whereas the MPS utilizes planning bills, the final assembly schedule will be based on product bills. The make-to-order environment will call for MPS planning at the component/raw material level. Again, the final assembly or finished product schedule is based on the product bill of material.

A basic rule to follow is that the MPS level should correspond to the desired planning and stocking level. Figure 4-1 illustrates the environmental, MPS, and final assembly relationships.

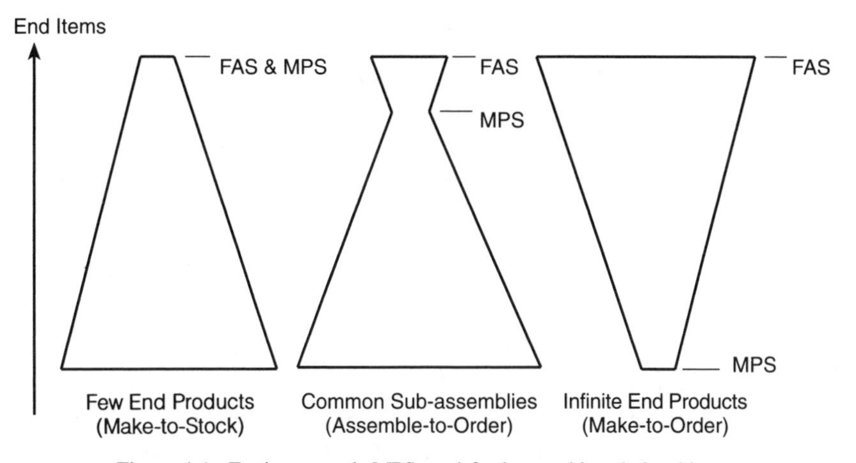

Figure 4-1. Environmental, MPS, and final assembly relationships.

Figure 4-2 compares the use of a traditional (product) structure with the modular structure of the same product line. Based on the multiple of combinations, 7680 different clocks can be produced. Master scheduling at the end item level utilizing the product structure would require forecasting and control of the 7680 items. Master scheduling at the modular level and with the modular structure would require forecasting and control of 44 items.

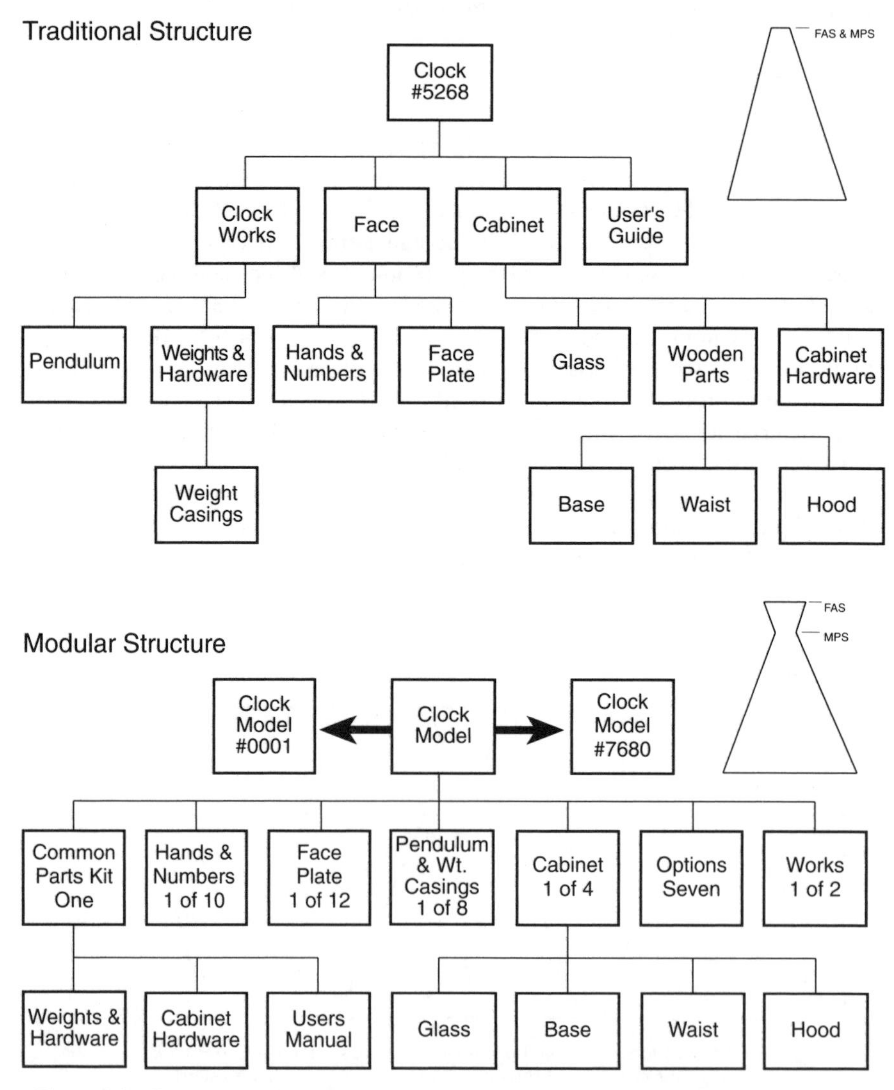

Figure 4-2. Comparison of traditional structure and modular structure. [Reprinted with the permission of APICS, Inc., from P.W. Stonebraker *Master Planning, Certification Review Course*, 1991.]

Final Assembly Applications

Whereas the MPS is an anticipated build or final product plan, the final assembly schedule (or finished product schedule) is the actual execution of the plan. It may be based on customer orders or stock replacement requirements. If the MPS bill of material structure is at the finished product level, the final assembly schedule (FAS) is compared easily to the MPS. If the MPS is based on planning bills utilizing modules, common parts bills, and/or components, the FAS must be coordinated with the MPS items. The FAS requirements must be compatible with the requirements generated from the MPS. An overstated FAS will create item shortages, whereas an understated FAS will cause readjusting the MPS and ultimately the MRP requirements.

Whereas the MPS may be based on planning bills, the FAS will require a single-level product bill for the purpose of component and subassembly order picking as well as product costing. The lead time of the FAS should be as short as possible and expressed in lesser periods, such as shifts or days. Final assembly products may be radically different, such as a four-door sedan and a convertible and still be produced on the same assembly line. Translating the output of the order entry system to the FAS product bill of material is often a challenge.

MANAGING THE MPS

Logic

The MPS plan is a time-phased record that compares forecasted activity, projected available inventory, and anticipated production (the MPS). Table 4-1 shows a 10-week horizon with a forecasted increasing sales rate (100-150 units in week 7), production lot sizes of 300, a beginning on-hand balance of 120, and projected on-hand inventories for each period. Note that the projected on-hand balance at the end of week 9 is negative, indicating the need for a plan adjustment.

Table 4-1. MPS Plan for Ten-Week Horizon; Lot Size 300, On-Hand 120.

	Weeks									
	1	2	3	4	5	6	7	8	9	10
Forecast	100	100	100	100	100	100	150	150	150	150
Projected on Hand	320	220	120	320	220	120	270	120	−30	120
MPS	300			300			300			300

Table 4-2 represents an MPS plan that includes customer orders. Note that in week 1, the customer orders were considered in the projected on-hand balance calculation. This is based on a management decision that the first demand week time fence is frozen and that only the customer order quantity will be considered. Beyond the one week time fence, a management decision is to calculate the projected on-hand balance by using the larger of either the forecast or customer orders.

Table 4-2. MPS Plan Including Orders; Lot Size 100, On-Hand 40.

	Weeks							
	1	2	3	4	5	6	7	8
Forecast	25	25	25	25	25	25	25	25
Orders	23	27	17	8				
Projected on Hand	17	90	65	40	15	90	65	40
MPS		100				100		

Although the time-phasing calculation is straightforward and the same as in MRP logic, it is recommended that the computer system should not control the maintenance of the MPS. Management control is continuously required to maintain a relatively stable and realistic plan. "Firming" MPS orders is a technique that will not allow the computer system to change the quantity or timing of the order.

Available-to-Promise

The available-to-promise (ATP) quantity calculation is shown in Table 4-3. It is the uncommitted quantity of projected on-hand inventory in each period when an MPS order receipt is planned. These quantities are useful in customer order promising. Note that the ATP is calculated for each period of time covered by the MPS order and is not cumulative. This is based on the assumption that although there are unsold units of 17 in week 1 and 48 unsold projected units in weeks 2 through 5, these will be sold by week 6 and, therefore, only the MPS order of 100 will be available at that time.

Time Fences

Implementing policy decisions relative to control of operating procedures over differing time periods is accomplished with time fences. The up-close period of the first week or two may be frozen with no changes allowed. The intermediate period going out to the total cumulative lead time for parts

Table 4-3. Available-to-Promise Quantity; Lot Size 100, On-Hand 40.

	Weeks							
	1	2	3	4	5	6	7	8
Forecast	25	25	25	25	25	25	25	25
Orders	23	27	17	8				
Projected on Hand	17	90	65	40	15	90	65	40
ATP	17	48				100		
MPS		100				100		

availability would have less stringent controls, and the period beyond the cumulative lead time would be unrestricted.

The actual time-fence decisions are dependent on the manufacturing environment. In a make-to-stock environment, the firm or frozen time period would include customer order processing, whereas the assembly and component lead time would be somewhat flexible (slushy). In the assemble-to-order environment, the firm period would include customer order lead time and assembly time with the component lead time slushy. The make-to-order firm period would include component lead time as well as assembly lead time. Common raw material procurement lead time will be within the slushy time period.

The firm time period is defined as being within the demand time fence, whereas the limited flexibility or slushy time period is beyond the demand time fence but within the planning time fence. Beyond the planning fence, the period is considered free and completely flexible and is primarily used for capacity planning.

The Master Scheduler

The responsibility for the MPS includes the entire organization: Marketing will input the forecast; engineering will structure the bills; finance will be responsible for approval of the required assets; and manufacturing will develop and execute the plan. The day-to-day development and maintenance of the MPS is the responsibility of the master scheduler.

Assuming that the product structure for the MPS has been established, the determination of control variables such as time fences and safety stock made, and an approved forecast in place, the initial capacity tests must be taken. Once satisfied that the plan is doable, the MPS orders must be firmed. Continual review of the MPS is required to assure the following:

1. The forecast is being consumed as planned.
2. The MPS orders are on schedule.
3. The capacity load remains realistic. Because one or possibly all three of the above do not always happen as planned, the master scheduler must have a complete understanding of the entire manufacturing system in order to do an intelligent analysis of the data for purposes of replanning. As stated earlier, communicating the real world to management is most critical.

CASE STUDY

Problems

The initial MPS run shows the information in Table 4-4.

Table 4-4. Initial MPS Run; Lot Size 40, On-Hand 4.

	Weeks											
	1	2	3	4	5	6	7	8	9	10	11	12
Forecast	10	10	10	10	10	10	10	20	20	20	20	20
Orders	10	12	8									
Projected on Hand	34	22	12	2	32	22	12	−8	12	−8	−20	−48
ATP	14				40				40			
MPS	40				40				40			

1. A customer requests your best promise on a new order for 20 items. He will accept split shipments. What do you promise?
2. What action should be taken to correct the projected negative balances?

Solutions

1. Promise the customer 14 units to ship out of week 1 production. The remaining 6 units should be promised from the scheduled MPS order of 40, which should be moved up from week 5 if possible. If not possible, promise in week 5.
2. If possible, the MPS order scheduled in week 9 should be moved to week 8. This action will cover the forecast of weeks 8 and 9 and part of week 10. A second order for 40 should be placed for delivery in week 10. A third order will be required for delivery in week 12.

QUIZ

1. The functions of the MPS over the short horizon are
   I. driving the MRP
   II. planning short-term capacity requirements
   III. recommending production of components
   IV. planning order priorities

   a. I, II, and III
   b. I, III, and IV
   c. I, II, and IV
   d. All of the above

2. The MPS should strive to
   I. maintain a balance between scheduled load and available productive capacity over the short term
   II. form basis for establishing planned capacity over the long term

   a. I
   b. II
   c. I and II
   d. Neither I nor II

3. Development and administration of the MPS is the responsibility of
   I. marketing
   II. manufacturing
   III. finance

   a. I and II
   b. I and III
   c. II and III
   d. All of the above

4. The MPS must maintain MRP realism between planning and
   a. engineering
   b. finance
   c. execution
   d. marketing

5. The master production schedule can supply the marketing department with information on the status of shipable end items for all of the following products EXCEPT
   a. dependent demand components
   b. single-model appliances
   c. consumer package goods
   d. custom-built machines

6. Exponential smoothing is a routine method for calculating
   a. safety stock
   b. forecasts

c. standard deviations
d. tracking signals

7. Forecast bias is measured with
   a. exponential smoothing
   b. demand filters
   c. standard deviations
   d. tracking signals

8. A greater degree of accuracy will be possible if the forecast is
   I. by product groups
   II. for a short term

   a. I
   b. II
   c. I and II
   d. Neither I nor II

9. Forecast error can be measured through
   I. standard deviations
   II. mean absolute deviations

   a. I
   b. II
   c. I and II
   d. Neither I nor II

10. Demand data entry error may be detected by
   a. a tracking signal
   b. a demand filter
   c. second-order smoothing
   d. a weighted average

BIBLIOGRAPHY

Fogarty, D.W., Blackstone, J.H., Jr., and Hoffman, T.R., *Production & Inventory Management,* 2nd ed. Cincinnati: South-Western Publishing, 1991.

Plossl, G.W., *Production and Inventory Control—Principles and Techniques*. Englewood Cliffs, NJ: Prentice-Hall, 1985.

Stonebraker, P.W., *Master Planning, Certification Review Course*. Falls Church, VA: American Production and Inventory Control Society, 1991.

Vollman, J.E., Berry, N.L., and Wybark, D.C., *Manufacturing and Control Systems,* 3rd ed. Homewood, IL: Richard D. Irwin, 1992.

5
Concepts and Logic of MRP

PURPOSE

Planning

Material requirements planning (MRP) is a tool used in the inventory management of dependent components in manufacturing operations. All items listed in a bill of material (purchased, fabricated, or subassembled) can be controlled through the MRP system. Notice the use of the word can. Some items such as hardware or packaging may be more easily controlled through alternative methods such as two-bin or reorder point but would still be listed in the bill of material for purposes of product costing and understanding. Software systems allow the user to code such items as "non-MRP" just as end items are coded as "MPS" rather than MRP.

Material requirements planning is a component fabrication or purchasing planning system which will recommend action calls at the appropriate time. Valid needs will be made known based on master production schedule (MPS)-driven requirements. The plan is dynamic in that every recorded data change such as inventory transactions, bill of material changes orders, and MPS revisions that occur between MRP generated runs will affect the plan.

Time Phasing

As stated earlier, a reorder point system only addresses a required quantity need at one point in time. An example would be an item with a lot size of 1000 and a calculated reorder point of 1300 based on a lead time of 7 weeks. When the on-hand and on-order quantity drops to 1300, the system calls for another order of 1000 to be released for delivery in 7 weeks. The system does not address the requirements beyond the 7-week delivery nor does it evaluate the timing of any open orders relative to the projected needs at scheduled receipt. If an item has a forecasted independent demand that is uniform over an extended time period, the reorder point system will work quite well.

When an item does not meet the criteria of uniform independent demand, time phasing—the stating of anticipated future demand and inventory planning by time periods—is required. Whereas the reorder point addresses quantity only, time phasing addresses both quantity and timing. The time-phased periods in most MRP systems are in weekly time buckets in which all data are accumulated in weekly periods. Based on the data input, requirements are, in turn, calculated and stated in weekly time periods.

Care must be taken to define and understand when an activity is taking place within the time period. The conventions are as follows:

1. All actual and planned replenishment orders are due at the beginning of the period. If the time period is weekly and starting on Sunday, any orders due during the previous week will show due at the beginning of the next week.
2. The anticipated usage will be the demand during the week.
3. The projected inventory will be at the end of the week.

Table 5-1. Partial Time-Phase Record

	Week	
	37	38
Gross requirements		100
Scheduled receipts		120
Projected inventory	40	60

Table 5-1 is an illustration of the required timing for which

- Week 38 starts Sunday, September 11, and ends Saturday, September 17.
- Anticipated gross requirements of 100 are from Monday, September 12, through Friday, September 16.
- Scheduled receipts totaling 120 are 70 on Wednesday, September 7, and Friday, September 9.
- The projected inventory at the end of week 37 is 40.
- The projected inventory at the end of week 38 is 60.

A bucketless system uses a daily time period but can then collect and display based on defined time periods such as weeks. Daily time periods can be used to advantage in greater planning and control in the execution of the plan.

Capacity Insensitive

The MRP system is a planning system which will list the fabrication and purchasing needs to meet the MPS. If the bills of material, the inventory data, and the planning factors are accurate, it will supply valid data. Although the plan may be valid, it does not mean that the plan is realistic. The MRP calculation of needs does not address capacity; in other words, it assumes infinite capacity. If a sudden forcast increase is reflected in the MPS for next month, the MRP system may call for increased fabrication activity that is well beyond the capacity of the plant and suppliers.

The maintenance of a realistic MRP plan is achieved through capacity management that is the constant monitoring and measuring of requirements and comparison to capacity limits. The correction of an unrealistic MRP plan can be achieved through adjusting the MPS not the MRP. The MRP will only list the requirements generated by the MPS. A successful plan must list valid requirements and be doable.

INPUTS AND OUTPUTS

The Master Production Schedule

The major input to the MRP system is the MPS or "driver" of the system. The manufacturing environment will determine the planning level of the product to be master scheduled. End items, major subassemblies, major components, or raw materials can be master scheduled. The minimum length of the MPS horizon for MRP purposes is the longest critical path of the master scheduled items. The horizon normally extends beyond the minimum required for material planning in order to plan capacity. Adjusting the MPS is often called for in order to reconcile capacity problems.

The Bill of Materials

The input of the parent-component relationship required for MRP is in the product structure or bill of material file. As explained earlier, the bills of material must be compatible with the items in the MPS, that is, product and/or planning bills. The bills must reflect all levels (multilevel) in order to plan all components.

The explosion of all MPS items through all levels in the bills of material will determine the requirements based on parent-component relationships. The system will combine the requirements of items that are common to more than one parent, and it will also combine requirements of items that occur at more than one level in the bills.

Inventory Status Data

Inventory status data are maintained in the item master or parts master file. The data are either planning factors which are static and are user maintained or inventory quantities which are considered dynamic and are transaction driven. Examples of planning factors required for MRP systems are as follows:

1. Lot Size Order Policy Rule. There are a variety of lot size rules that would be applicable to different items within the system. Consideration mut be given to setup costs, quantity discounts, carrying costs, usage variation over the horizon, and so forth
2. Safety Stock. Planned safety stocks may be based on quantity or safety lead time.
3. Manufactured or purchased.
4. MRP, MPS, or non-MRP controlled.
5. Unit of measure.
6. Planning group or family.
7. Low-Level Code. This is the lowest level which the component is listed in any bills of material. This code is useful in the calculation of requirements for multilevel items.
8. Planned Lead Time. This is the estimated overall lead time for both manufactured and purchased items. It is used for lead-time offset in the MRP calculation. It is an estimate and will not exactly match the sum of the calculated operational lead times when scheduling the shop order based on the routing file. Actual lead times can be affected by shop work loads and varying order quantities. Although not totally accurate, the two lead times must be reasonably close.

Inventory data in the item master file are both quantity on-hand and on-order. On-order information is based on either open shop or purchase orders. A record of allocated stock is also maintained in an allocation or material requirements file. This allocated or reserved stock is planned for released shop orders, but has yet to be taken from stock. As it is not available for other planning, the allocated quantity must be subtracted from the on-hand quantity. When the material is withdrawn from stock, both the on-hand and allocated quantities must be reduced.

Planned Order Output

A planned order is based on the net requirement of an item adjusted for lot size and with a release date based on lead time offset from the required due date. Planned order due dates extend the planning horizon of the MRP system. Planned orders are calculated within the system and will be adjusted or canceled when the system is regenerated. A release date in the current week is a recommendation to initiate the purchase or manufacturing order of the item. The term manufacturing order is used interchangeably with stop order.

Planned orders within the system with release dates in the future project requirements for the next level down. An example would be a planned order to produce a subassembly starting in 6 weeks. The gross requirements for the subassembly components will be listed in week 6. Future planned orders also serve as inputs to capacity planning systems in order to calculate future work-load profiles.

Recommended Action Notices

In addition to the recommendation to release current planned orders, the system will make recommendations to expedite (move up), deexpedite (move back), or cancel existing orders. Prior to MRP, the first notice to expedite a part was often when there was a shortage. There was no available information to tell the planners that an item could be moved back on the schedule. With each regeneration of the MRP system, recommendations are made based on forecast changes, inventory adjustments, scrap or rework, unplanned shipments, and so forth.

Although the system is sensitive to change, at times it may be too sensitive and create a nervous system which results in shop or purchases orders being "jerked around" needlessly. The system can be stabilized through the use of "firmed planned orders" which will be explained in the chapter devoted to managing the MRP.

GROSS TO NET REQUIREMENTS

Projected Inventory Status

The time-phased inventory status is the heart of the MRP calculation. Table 5-2 represents a 10-week period of projected activity. The gross requirement of 100 per week is the anticipated demand in each time period. The on-hand quantity going into week 1 is 100. The scheduled receipt of 300 due by the beginning of week 2 is based on an open purchase or shop order. The projected on hand for each week is equal to the projected on hand at the beginning of the week plus any scheduled receipts for the week minus the gross requirement projected during the week. The projected on hand at the end of week 2 is

$$0 + 300 - 100 = 200$$

Note that the only scheduled receipt is in week 2 so that when the projected on hand goes negative in week 5, the negative balance accumulates for the balance of the weeks.

The net requirement is a projected shortage for each time period. It is the gross requirements for the time period subtracted from the projected beginning

Table 5-2. Projected Activity; Lot Size = 300, Lead Time = 2 Weeks

	Weeks									
	1	2	3	4	5	6	7	8	9	10
Gross requirements	100	100	100	100	100	100	100	100	100	100
Scheduled receipts		300								
Projected on Hand	0	200	100	0	−100	−200	−300	−400	−500	−600
Net requirements					100	100	100	100	100	100
Planned order receipts					300			300		
Planned order release			300			300				
Projected available balance	0	200	100	0	200	100	0	200	100	0

on hand and the scheduled receipts for the week. If the result is positive (i.e., the gross requirement is covered), there is no net requirement. Once the projected on-hand inventory runs out (goes negative), the net requirements for each time period equals the gross requirement for that time period. Weeks 5 through 10 all indicate both gross and net requirements of 100 for each week.

The planned order receipt is calculated to cover net requirements. It will cover the net requirements with an order quantity based on its specific lot size rules. In Table 5-2, there is a planned order receipt of 300 in week 5 to cover the net requirements for weeks 5, 6, and 7. A second planned order receipt is due in week 8. The releases for the planned orders are lead time offset by 2 weeks as shown by the planned releases in weeks 3 and 6. The planned order release date is most critical in the linking of the MRP records and will be explained in detail later in this chapter.

The accumulating of the negative projected on-hand balances is known as the negative balance method. The reasoning is that only scheduled receipts; that is, released orders should be used in the projection. Planned orders are not released and are subject to recalculation with each regeneration of the system. An exception to recalculation is with a net change system and is explained later in this chapter. The accumulated negative balances can be confusing, and therefore a projected available balance calculation has been added. The projected available balance considers both scheduled and planned receipts. Note that in the first 4 weeks in Table 5-2 that the projected available balance considers the planned orders.

Shop Calendar

A shop or manufacturing calendar is used for the time-phased calculations required with MRP systems. It consecutively numbers the predetermined

Table 5-3. Shop Calendar (June–July).

SUN	MON	TUES	WED	THURS	FRI	SAT
29	30	1	2	3	4	5
	422	423	424	425		
6	7	8	9	10	11	12
	426	427	428	429	430	
13	14	15	16	17	18	19
	431	432	433	434	435	

working days in the manufacturing operation. The calendar converts manufacturing days into Gregorian calendar days and can be grouped or displayed in desired time periods such as weeks. This numbered-day approach has the advantage of simplified arithmetic, as well as the ability to plan and execute lead times with a tighter degree of control. Table 5-3 is an example of a portion of a shop calendar. Note that Saturdays and Sundays, as well as July 4, are not scheduled (nor numbered) work days. If the user has defined weekly planning time periods, the availability of a item is assumed at the beginning of the week. An order to be in stock the week of July 6 must be scheduled for receipt prior to day 426.

If a customer order for an item not in stock is required on July 16 (day 433), it is a net requirement for the week of July 13. Therefore, the item is planned to be in stock by day 431 (the beginning of the week). If the lead time is 7 days, the release date will be

431 - 7 = Day 424 = Week of June 29

Using daily execution control, the order could be released at the beginning of the week of July 6:

433 - 7 = Day 426 = Week of July 6

Lead-Time Calculation

The rules for the planning lead time to be used for required receipt offset to planned release date are as follows:

1. The planning lead time is to be maintained in the item master file in the same time increment as the planning time period. In the above example where weekly planning periods are used, the 7-day lead time would be stated as 2 weeks. The release week would be the week of July 13 minus 2 weeks = week of June 29.

2. The planning lead time is to include all time elements such as review period, supplier lead time, and inspection and stocking time for a purchased item. A manufactured item must include estimates of queue (waiting) times, run, and setup and move times.
3. The planning lead time is a estimate for the total lead time and normally will not exactly match the sum of the individually calculated operation-by-operation times generated by the system for shop order dispatching. The difference can be due to lot size differences, individual work center changes in waiting time allowances, and the rounding out to weekly time periods. Although the planning lead time will not exactly match the dispatched or scheduled times, it should be relatively close. In most situations, the difference between planned and scheduled lead time is less of a problem than the difference between actual and planned lead time.

EXPLOSIONS

The requirements explosion is the linkage of parent to component relationships through all levels in the bill of material. Prior to consideration of lead time offset or lot sizing, the Level "0" orders in the master schedule drive the gross requirements of those Level 1 components required for the parent. The net requirements for the Level 1 components are then calculated as shown in Table 5-2 (net requirements = on-hand + scheduled receipts − gross requirements). The net requirements of Level 1 items explode the bill of materials to create the gross requirements of Level 2 items. This logic follows through all levels of the bill of material.

Based on the following bill of material, the gross to net level-by-level calculation is shown in Table 5-4. This parent-component linkage reflects the gross-net relationship but does not allow for the lead-time offsets or lot sizing.

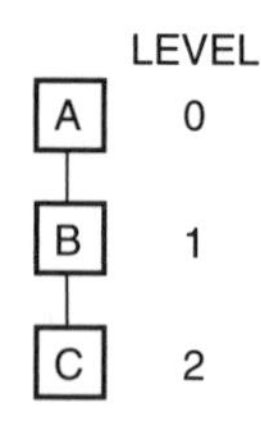

Lead-Time Offsets

Table 5-5 reflects the gross-net relationships shown in Table 5-4 but offsets the gross requirement time period of the component to allow for the required lead time.

Table 5-4. Planning Grid—No Lead-Time Offset nor Lot Size Consideration

Part	Time Periods						
	0	1	2	3	4	5	6
A							
Master schedule				10		10	10
B							
Gross requirements				10		10	10
Scheduled receipts							
Projected on hand	12	12	12	2	2	−8	−18
Net requirements						8	10
C							
Gross requirements						8	10
Scheduled receipts						5	
Projected on hand						−3	−13
Net requirements						3	10

To produce the parent or at the lowest level to purchase the component (the lowest level in a parent-component chain is always purchased). The lead time for part A is 2 weeks, whereas that for parts B and C is 1 week. As shown, the planned order receipt meets the needs reflected by the net requirement and the planned order release offsets the lead time to allow for receipt in the planned time period.

Lot Sizing

Planned orders must reflect the lot size rules assigned to each item. If the lot size rules for items B and C were lot for lot, planned orders would be as shown in Table 5-5. However, if setup costs, supplier price breaks, or other factors affecting lot sizing causes lot sizes greater than the single time period requirements, those lot sizes will be reflected in the planned orders. Assuming a lot size of 12 for item B and 5 for item C, the relationships are reflected in Table 5-6. Note that a projected available balance listing has been added.

After consideration for lead-time offset and lot sizing, the planned order release for a parent in a given period causes a gross requirement of the component in that period. In other words, to start to produce an item, the components must be there.

Table 5-5. Planning Grid with Lead-Time Offset

Part		Lead Time	Time Periods						
			0	1	2	3	4	5	6
A		2							
Master Schedule						10		10	10
B	Lot for Lot	1							
Gross requirements				10		10	10		
Scheduled receipts									
Projected on hand			12	2	2	−8	−18	−18	−18
Net requirements						8	10		
Planned order receipt						8	10		
Planned order release					8	10			
C		1							
Gross requirements					8	10			
Scheduled receipts					5				
Projected on hand			0	0	−3	−13	−13	−13	−13
Net requirements					3	10			
Planned order receipt					3	10			
Planned order release				3	10				

Time-Phased Order Point

The time-phased order point is a technique used for planning independent demand of items with gross requirements either forecasted or customer order controlled. Time-phased order point controlled items would include end items reflected in the master schedule as well as service requirements of items that may have dependent, as well as independent, demand. If forecasted demand is uniform and continuous, the conventional order point will work out; but if requirements are anticipated to be nonuniform (lumpy), the time-phased reorder point system (where, in fact, there is not an order point) is called for. The gross-net requirements logic of MRP is utilized with allowances made for lead time and lot sizing. Table 5-2 shown earlier in this chapter could apply to either a time-phased order point controlled master schedule or an MRP controlled item. The difference is in the management of the system.

Regenerative and Net Change Systems

There are two methods used in the approach to the replanning cycle of MRP. They differ in the timing and the data utilized in the process, but the logic and the output remain the same. The regenerative system is time-driven and

Table 5-6. Planning Grid with Lead-Time Offset and Lot Size Consideration

Part	Lot Size	Lead Time	Time Periods						
			0	1	2	3	4	5	6
A		2							
Master Schedule						10		10	10
B	12	1							
Gross Requirements				10		10	10		
Scheduled receipts									
Projected on hand			12	2	2	−8	−18	−18	−18
Net requirements						8	10		
Planned order receipt						12	12		
Planned order release					12	12			
Projected available balance			12	2	2	4	6	6	6
C	5	1							
Gross requirements					12	12			
Scheduled receipts					5				
Projected on hand			0	0	−7	−19	−19	−19	−19
Net requirements					7	12			
Planned order receipt					10[a]	10[a]			
Planned order release				10	10				
Projected available balance			0	0	3	1	1	1	1

[a]Two orders of 5 are required for each period.

normally scheduled to be run weekly. It calls for a total explosion of all bills of material for master scheduled items. All planned orders from the previous explosion are recalculated. Although there may be a comfort level in the disciplined approach to a weekly time cycle, there are disadvantages of which to be aware, such as deterioration of the plan as the week progresses and a very heavy load of output with the weekly explosion.

The net change system calls for a partial explosion and will relate only to those items affected by a change since the last explosion. The system is transaction-driven rather than by time and is often processed daily or in some situations on a real-time, on-line basis. With net change, the planned orders and their associated requirements are not erased but will be rebalanced if a transaction has affected the inventory, bill of material, or requirements. Net change has the advantage of being more responsive to change, is continually up to date, and evens out the work load due to less output per run. Care must be taken in that a responsive system may become a nervous system causing never-ending replanning and rescheduling of requirements. Partial control of nervous output can be achieved by "firming" planned orders as mentioned earlier. This technique will be discussed in the next chapter on managing MRP. If the activity levels of finished goods and component service parts are such that almost all items are affected daily, the net change system will call for large computer resources and planning activities. It will be similar to a regenerative system that regenerates daily.

The decision to operate in a regeneration or net change mode will be dependent on the nature of the manufacturing environment. It should be remembered that MRP is for planning and that timely execution might best be controlled by an off-line technique that would be compatible with the MRP plan.

IMPLOSIONS

Imploding is the reverse of the explosion process and it determines the where-used relationship in a bill of material. It can be for any single-level relationship or it may start at the lowest level of the bill and implode up the bill to the end item level. A full implosion of an item will show all parent-component relationships of that item in the bill of material file.

Pegging

Pegging is the use of the where-used implosion technique to identify a single source or sources for the gross requirements of an item. Level-by-level pegging will trace the source through all levels of the bill up to the original source, the master scheduled item. Unlike a where-used full implosion, pegging will only show those relationships that produce the gross requirements of the item. The key to understanding will be the determination of the planned order release of the parent that, in turn, creates the gross requirement of the component.

Referring back to Table 5-6, purchased part C calls for a week 1 release of 10 units for delivery in week 2. If the planner is told that there will only be 5 units available for delivery in week 2, level-by-level pegging will reveal the impact of the problem in the following manner.

PART	PERIOD	QUANTITY REQUIRED	SOURCE
C	2	12	Part B
C	3	12	Part B
B	3	10	Part A
B	4	10	Part A

With this information, the planner would then call up the detailed information shown in Table 5-6. Analysis of the data would show the following:

Table 5-7. Replan After Pegging

Part	Lot Size	Lead Time	Time Periods						
			0	1	2	3	4	5	6
A		2							
Master Schedule						10		10	10
B	12	1							
Gross requirements				10		10	10		
Scheduled receipts									
Projected on hand			12	2	2	−8	−18	−18	−18
Net requirements						8	10		
Planned order receipt						8	12		
Planned order release					8	12			
Projected available balance			12	2	2	0	2	2	2
C	5	1							
Gross requirements					8	12			
Scheduled receipts					5				
Projected on hand			0	0	−3	−15	−15	−15	−15
Net requirements					3	12			
Planned order receipt					5	15			
Planned order release				10	10				
Projected available balance			0	0	2	5	5	5	5

1. To meet the week 3, part A schedule, only 8 B parts must be fabricated, not the desired lot size of 12.
2. To meet the adjusted planned releast of 8 B parts in week 2, only 3 more will be required, not the 10 originally planned.
3. If the supplier can "catch up" and deliver 15 in week 3, 12 B parts can be produced and a projected available balance in week 4 will be 2. See Table 5-7.
4. If the supplier cannot "catch up" and can only deliver 10 in week 3, 10 B parts can be fabricated, not the desired 12, but the total requirements of item A in the master schedule will be met.

CASE STUDY

Problem

A subcontractor supplies customer A with machined gears. The lead time to machine the gears is 2 weeks, and they are machined lot for lot. The gear blanks are purchased in lots of 200 with a lead time of 4 weeks. Customer A has supplied a time-phased forecast as follows:

Week	Gears
Week 1	100 Gears
3	120
5	120
7	150
9	160

There are 100 machined gears scheduled for receipt in week 1 and 120 in week 3. There is a scheduled receipt of 200 gear blanks in week 3.

1. When should the subcontractor plan to release additional shop orders and in what quantities?
2. The gear blank supplier has requested a forecast of orders to be released in the next 3 weeks. What is the forecast?
3. How many gear blanks are projected to be on hand going into week 8?

Solution

Time-phased analysis of the parent-component data is as follows:

	Week									
	1	2	3	4	5	6	7	8	9	10
Gear										
Forecast	100		120		120		150		160	
Scheduled receipts	100		120							
Net requirements					120		150		160	
Planned order receipts					120		150		160	
Planned order release			120		150		160			
Gear blank										
Gross requirements			120		150		160			
Schedule receipts			200							
Projected on hand			80		−70		−230			
Net requirement					70		160			
Planned order receipt					200		200			
Planned order release	200		200							
Projected available balance			80		130		170			

1. Planned order fabrication releases are
   Week 3—120
   Week 5—150
   Week 7—160
2. The next 3-week gear blank purchase order release forecast is
   Week 1—200
   Week 3—200
3. The projected available balance of gear blanks going into week 8 is 170.

QUIZ

1. MRP evolved from combining the following two principles:
   I. Dependent demand calculation
   II. Independent demand calculation
   III. Time phasing
   IV. Inventory accuracy

   a. I and II
   b. I and III
   c. II and III
   d. II and IV

2. The time-phased order point
   I. is a technique for controlling independent demand items
   II. has system processing logic similar to MRP
   III. is suited for service parts
   IV. ignores the aspect of specific timing

   a. II
   b. I and II
   c. I, II, and III
   d. All of the above

3. Prerequisites for an MRP system are:
   I. master production schedule
   II. bill of material
   III. unique part numbers
   IV. available inventory numbers

   a. I, II, and III
   b. I, II, and IV
   c. II, III, and IV
   d. All of the above

4. An MRP system answers the following:
   I. What material is needed?
   II. In what quantities?
   III. When is it needed?

a. I and II
b. I and III
c. II and III
d. All of the above

5. "Netting" consists of allocating against "gross"
   I. on-hand inventory
   II. on order

   a. I
   b. II
   c. I and II
   d. Neither I nor II

6. The planning horizon
   I. is related to the longest cumulative procurement and manufacturing lead times for components
   II. can be less than the longest cumulative lead time

   a. I
   b. II
   c. I and II
   d. Neither I nor II

7. A numbered-day shop calendar
   I. considers only scheduled working days
   II. makes scheduling arithmetic straightforward

   a. I
   d. II
   c. I and II
   d. Neither I nor II

8. The lead time of a manufactured item includes
   I. run time
   II. setup time
   III. queue time
   IV. wait time

   a. I and II
   b. I and III
   c. I, II, and IV
   d. All of the above

9. Regenerative MRP requires
   I. every MPS item must be exploded
   II. every active bill of material must be retrieved
   III. the status of every active item must be recomputed
   IV. voluminous output may be generated

   a. I, II, and III
   b. I, II, and IV
   c. II, III, and IV
   d. All of the above

10. The main difference between regenerative and net change MPR is
    I. the frequency of replanning
    II. what sets off the replanning process

    a. I
    b. II
    c. I and II
    d. Neither I nor II

BIBLIOGRAPHY

APICS Dictionary, 7th ed. Falls Church, VA: American Production and Inventory Control Society, 1992.

Lunn, T. with Neff, S.A., *MRP—Integrating Material Requirements Planning and Modern Business*. Homewood, IL: Business One Irwin, 1992.

Orlicky, J., *Material Requirements Planning*. New York: McGraw-Hill, 1975.

St. John, R.E., *Material and Capacity Requirements Planning Certification Review Course*. Falls Church, VA: American Production and Inventory Control Society, 1991.

Vollman, J.E., Berry, N.L., and Wybark, D.C., *Manufacturing and Control Systems,* 3rd ed. Homewood, IL: Richard D. Irwin, 1992.

6
Managing MRP

Once the basic logic of MRP (time phasing and parent–component relationships) is in place and understood, there are planning factors and techniques that must be utilized in the management of the system. The requirements generated by the software utilizing gross-net logic will always be mathematically correct, but their usefulness will be dependent on realistic control numbers such as lot sizes and safety stocks as well as the management of the requirements through releasing and rescheduling.

LOT SIZING

Every MRP managed part must have a lot size rule. This is true for manufactured or purchased parts. It has been long understood that large lot sizes would compensate for big setup costs. It was also understood that although there was the advantage of the large lot sizes absorbing the setup costs, there was the disadvantage of the costs of carrying the inventory. In the early part of the century, a formula was developed that balanced these conflicting factors and thus supplied an economic order quantity (EOQ). This was the start of studies, restudies, analysis, and reanalysis of lot size management. It is felt by many, including this author, that too much emphasis was put on lot-sizing rules to the detriment of other materials/logistics management control philosophies such as lead time and capacity management.

In the following review of the various lot-sizing techniques, note that the variables are setup costs, anticipated volume, and the inventory carrying costs. The volume or rate of usage is based on forecasts that are always to some extent wrong. The inventory-carrying cost is an estimate which can have a large range depending on whether cost of borrowing money or an opportunity cost factor is considered. The Japanese taught us that setup cost should not be considered fixed but should be reduced. Table 6-1 illustrates the differences in calculated lot sizes using the same economic order quantity formula for a similar situation but adjusting the variables. The setup cost in Alternative 1 is \$100, but is reduced to \$80 in Alternative 2—a realistic result

Table 6-1. Lot Size Comparisons

	Alternative 1	Alternative 2
Annual usage	6000 Units	6000 Units
Standard unit cost	$2.00	$2.00
Setup cost	$100	$80
Inventory-carrying cost	16%	32%
Calculated lot size	1937 Units	1225 Units

of setup reduction. In both alternatives, a storage, handling, and obsolescence estimate of 6% is used; but in Alternate 1, a cost of money allowance of 10% is used, bringing the carrying cost of Alternate 1 to 16%. In Alternate 2, an opportunity cost of 26% is added to the storage, handling, and obsolescence estimate, making the carrying cost 32%.

In the calculation of purchased parts' lot sizes, the setup cost is often replaced with the cost of placing and receiving the purchase order. This calculation is valid for commodities that the supplier may have readily available. However, if the purchased part is an item produced for the customer's order, the real setup cost is the setup cost of the supplier. The supplier will often attempt to minimize his lot size cost by offering price breaks at differing order quantities. This will require separate analysis on the part of the customer.

The following three techniques do not directly consider setup and carrying costs.

The Fixed Period Requirement

A specified number of periods' requirements are ordered and the ordering interval is equal to the number of fixed periods. The lot size equals the sum of the fixed periods' requirements. For example, 3-week requirements would be ordered every 3 weeks. Table 6-2 illustrates a 2-week fixed period requirement reflected as planned order receipts.

Table 6-2. A 2-Week Fixed Period Requirement and Planned Order Receipts

	Week									
	1	2	3	4	5	6	7	8	9	10
Net	50	52	40	37	48	0	56	52	48	52
Planned order receipt	102		77		48		108		100	

The Fixed Order Quantity

A specific order quantity other than an EOQ quantity is based on a lot size logic other than a setup/carrying cost algorithm. It may be due to other factors such as die life for a manufactured part or price break for a purchased part. Table 6-3 shows the same net requirements as Table 6-2, but the planned order receipts are based on a fixed order quantity of 100.

By calculating the projected available balance, the reader will note that the fixed period requirement uses up the receipts every 2 weeks, but the fixed order planned order receipt had 65 units left over going into week 11.

Table 6-3. A Fixed Order Quantity of 100 and Planned Order Receipts

	Week									
	1	2	3	4	5	6	7	8	9	10
Net	50	52	40	37	48	0	56	52	48	52
Planned order receipt	100	100			100			100		100

Lot for Lot

This technique orders only what is required for each period. This is the most desirable technique not only minimizing inventory investment but also allowing reduced lead times and manufacturing flexibility. It is consistent with the Just-In-Time goals of lot sizes of one and setup costs of zero. Table 6-4 illustrates lot-for-lot ordering.

Lot-for-lot achieves the MRP goal of components being immediately consumed upon receipt. The remaining techniques consider the costs of setup or ordering as well as carrying costs.

Table 6-4. Lot-for-Lot Ordering

	Week									
	1	2	3	4	5	6	7	8	9	10
Net	50	52	40	37	48	0	56	52	48	52
Planned order receipt	50	52	40	37	48	0	56	52	48	52

The Economic Order Quantity

The economic order quantity (EOQ) quantity is based on the formula that balances setup or ordering costs with the cost of carrying inventory. It is:

$$\text{EOQ} = \sqrt{\frac{2 \text{ Annual demand} \times \text{setup cost}}{\text{Carrying cost percent} \times \text{unit cost}}}$$

The validity of the EOQ is based not only on the variables listed earlier in this chapter, but the assumption that the requirements are uniform and continuous. Table 6-5 illustrated the planned receipts when the EOQ is calculated to be 150.

Table 6-5. Planned Receipts for EOQ = 150

	Week									
	1	2	3	4	5	6	7	8	9	10
Net requirements	50	52	40	37	48	0	56	52	48	52
Planned order receipt	150			150				150		

The Period Order Quantity

The period order quantity is a calculation that evolves into a fixed period requirement based on a modified EOQ. It adjusts for the reality that the net requirements may not be uniform and continuous. For ease of explanation, Table 6-6 shows net requirements in monthly intervals. Based on the annual usage of 1200, an EOQ is calculated to be 200 and the irregular planned order receipts are shown to average out the receipts into a uniform pattern; the determination is as follows:

$$\frac{\text{Annual demand}}{\text{EOQ}} = \frac{1200}{200} = 6 \text{ Expected orders per year}$$

6 Orders per year = Ordering interval or fixed period of 2 months

Table 6-6. Net Requirements in Monthly Intervals and Period Order Quantity Planned Receipts

	Month											
	1	2	3	4	5	6	7	8	9	10	11	12
Net requirements	50	70	90	120	140	150	80	60	60	100	160	120
Planned order receipts EOQ calculation	200		200		200	200			200		200	
Planned order receipts period order quantity	120		210		290		140		160		280	

The planned order receipts generated by the period order quantity technique are shown in Table 6-6.

The Least Unit Cost, Least Total Cost, Part-Period Balancing, and the Wagner-Whitin Algorithm

These techniques all take into account the setup cost, the inventory-carrying costs, and the anticipated nonuniform rate of requirements. (If the requirements are uniform, the EOQ technique would be used). The techniques are as follows:

A. Least Unit Cost. Adds setup and carrying costs for each trial lot size and divides by the number of units in the lot size, picking the lot size with the lowest unit cost.
B. Least Total Cost. Compares the setup costs and the carrying costs for various lot sizes and selects the lot size where these costs are most nearly equal.
C. Part-Period Balancing. Uses same logic as least total cost, but adds a look-ahead/look-back feature to reevaluate each lot size.
D. Wagner-Whitin Algorithm looks at all possible combinations to pick the optimum.

Table 6-7 illustrates the results of two of the above four techniques.

Table 6-7. Planned Order Receipts Based on Least Unit Cost and Least Total Cost Techniques.

	Week									
	1	2	3	4	5	6	7	8	9	10
Net requirements	50	52	40	37	48	0	56	52	48	52
Planned order receipts (least unit cost)	102		125				108		100	
Planned order receipts (least total cost)	142			141				152		

Economic Order Quantity with No Instantaneous Receipt

The previously detailed lot-sizing techniques assumed that the lot size would be completed as a total lot at one time—instantaneous receipt—such as when a purchased order is received. In some manufacturing processes, the material may be received during a sustained period, such as a process line taking 1 week to complete 4 weeks of requirements. The noninstantaneous receipt EOQ formula is a modification of the standard EOQ formula:

$$\text{Noninstantaneous EOQ} = \sqrt{\frac{2 \text{ Annual Demand} \times \text{Setup Cost}}{(\text{Carrying Cost Percent} \times \text{Unit Cost})\left(1 - \frac{\text{Usage Rate}}{\text{Production Rate}}\right)}}$$

Table 6-8. Standard EOQ Versus Noninstantaneous Receipt

	Instant Receipt	Noninstant Receipt
Annual usage	6000 Units	6000 Units
Standard unit cost	$2.00	$2.00
Setup cost	$80	$80
Inventory-carrying cost	32%	32%
Production rate	Instant receipt	360 units/week
Usage rate	120 Units/week	120 Units/week
Calculated lot size	1225 Units	1500 Units

Table 6-8 compares Alternate 2 shown in Table 6-1 using the standard EOQ with the noninstantaneous EOQ based on a production rate of 360 units per week.

The EOQ, noninstantaneous EOQ, and the fixed order quantity techniques are considered nondiscrete in that their calculations allow for remnants of planned receipts to be carried over and not spoken for by net requirements. All other techniques are discrete in that the planned receipts (lot sizes) are all consumed within the planned period.

SAFETY STOCK

Safety stock, sometimes referred to as buffer or reserve stock, is planned stock to be available to handle fluctuations in demand and/or supply. Like all inventory, it does have a cost and therefore should be understood and used with discretion. Safety stock can be planned in units or in the form of safety lead time. Which method to use is dependent on an understanding of the cause of the fluctuation and the nature of the supply process.

Fluctuations in demand are the result of forecast deviations, wheras supply fluctuation can be due to supplier or manufacturing deviations from planned lead times or expected quantities.

Table 6-9. Example of Safety Lead-Time Control

Supply lead time = 3 weeks
Safety time = 1 week
Planning lead time = 4 weeks
Lot size = 40

	Weeks						
	1	2	3	4	5	6	7
Net requirements					20	20	30
Planned order receipts					40		40
Planned order release	40		40				

Safety Quantities

Planned safety stock quantities to allow for forecast deviations are maintained at the master production schedule level. The desired quantities are a function of deviations as explained in Chapter 4. There are some who feel that safety stock should not be carried at the intermediate levels of the bill of material. An exception to this approach would be component parts which have independent service usage and are therefore subject to forecast variation. Another exception would be when due to quality or process considerations, the process yield is somewhat unpredictable and therefore the manufactured or purchased quantity may be less than planned. A third exception for consideration is when modules and subassemblies are common to multiple finished products. Carrying safety stock at the subassembly rather than the master schedule level in this situation will allow greater flexibility with less inventory investment.

Safety Lead Time

Safety lead time is the adding of extra time to the planned lead time to allow for late delivery of either purchased or manufactured items. The amount of safety lead time should be based on the item's history of variation (lateness). Safety lead time is more practical than safety stock for dependent demand items, in that the time-phasing adjustment is not in conflict but is still consistent with planned order quantity requirements. If a supply order is late, a safety stock of 200 will not help if the net requirement is 320. Safety lead time should be maintained as a separate number, and although a part of the planned lead time, it should not be built into the supplier or shop order due dates. Table 6-9 is an example of safety lead-time control.

The planned receipt in week 5 will be based on a purchased or shop order due in week 4, while the planned receipt in week 7 will be based on an order due in week 6.

Table 6-10. Allocation of Safety Stock

On-hand = 30
Safety stock = 20
Lead time = 3 weeks
Lot size = 40

	Week							
	0	1	2	3	4	5	6	7
Gross requirement						30	20	30
Projected on hand	30[a]	10	10	10	10	-20	-40	-70
Net requirements						20	20	30
Planned order receipt						40		40
Planned order release			40		40			
Projected available balance		10	10	10	10	20	0	10

[a]Reduced from 30 to 10 to allocate safety stock

Table 6-11. Projected Available Balance Control

	Week							
	0	1	2	3	4	5	6	7
Gross Requirement						30	20	30
Projected on hand	30	30	30	30	30	0	-20	-30
Net requirements							20	30
Planned order receipt						40		40
Planned order release			40		40			
Projected available balance		30	30	30	30	40	20	30

Managing Safety Stock

Material Requirements Planning order control utilizing safety stock may be accomplished one of two ways: (1) treating the safety stock quantity as an allocation or (2) not letting the projected available balance drop below the safety stock quantity. Tables 6-10 and 6-11 illustrate the two methods.

To summarize, the best method of safety control would be to use safety stock for uncertain demand and safety time for uncertain lead time. In either case, it is critical that the planner/analyst understand how the safety stock or time fits into the planning system. When an expedite recommendation is received, analysis through pegging may show that the safety stock may cover the requirement or the order may be late and still meet the requirement. Safety stock should not be a crutch for poor performance, but it should be utilized when needed and expediting should be used to cover requirements, not to cover safety stock.

ALLOCATION

Allocated quantities are materials reserved within the MRP system. MRP logic is that a scheduled receipt does not consider component requirements in the same manner that a planned order receipt does. This logic is based on the assumption that the components have been issued or they have been allocated (taken from projected on-hand stock) at the time of order release. In other words, allocated stock is material on hand or on order assigned to a specific released order. The routine of allocating components at the time of order release is necessary, as picking orders does take time, and at any given time there will be "uncashed" requisitions that must be taken into account. The allocated quantity is subtracted from the projected on-hand quantity.

There may be situations when it is desirable to allocate future orders, such as a customer order for a service part component. This calls for time-phased allocation where the order is considered an additional requirement. Table 6-12 reflects an immediate allocation 20 and time-phased allocations in weeks 4 and 6.

Table 6-12. Planning Grid With Time-Phased Allocations

	Week						
	0	1	2	3	4	5	6
Gross requirements		20	20	20	20	20	20
Scheduled receipts			40				
Allocations	20				5		10
Projected on hand	50	10	30	10	-15	-35	-65
Net requirements					15	20	30

PLANNED ORDERS

Planned orders generated by MRP list suggested quantities, release, and receipt dates necessary to cover net requirements. The quantity is based on the predetermined lot size rule, and the release date is lead-time offset of the planned receipt date. The planned order differs from a scheduled receipt in that the former is a plan and the latter is a released purchase or shop order.

The recommendation to place a purchase or shop order based on a planned order release does not take place until the current period, also called the action bucket.

A planned order is created in the computer and will be detected and recalculated when the system is regenerated. In a net change system, planned orders are maintained and rebalanced when necessary. In either case, the planned order release explodes down to the next level and becomes the gross requirement of all its components. The planned order is also input to the capacity requirements system to predict future work loads by work center and time period.

Firm Planned Orders

The firm planned order (FPO) is a technique to stabilize the quantity and timing of a planned order. It overrides the normal logic of MRP for the purposes of overcoming nervousness or for a desired deviation from system logic, such as a one-time lot size rule change. Firming the planned order of a parent will stabilize all lower-level components. The greatest degree of stabilization is achieved by firming at the MPS level. All orders within a frozen time fence are, in effect, firmed.

Although an order is firmed, the computer will still recommend changes based on MRP logic. The farther out the horizon and the more orders firmed, the more recommendations with which to deal. Often the changing requirements can be covered through safety stock absorption or lead-time compression, rather than by jerking the system around.

Table 6-13 illustrates a conventional plan which reacts to a gross requirement change. Parent A calls for a safety stock of 15 and a lot size of 30. Component B plans for no safety stock and a lot size of 100.

The result of the gross requirement change of Parent A shifted the planned order release in week 4 to week 3. This, in turn, shifted the release of Component B from week 1 to "0"—in other words, less than planned lead time.

If the planned orders for Parent A had been firmed for weeks 4 and 6 receipt, the safety stock would have covered the increased requirement and the planned order receipt for week 7 would have restored the safety stock. This is shown in Table 6-14.

RELEASING ORDERS

The releasing or launching of orders is recommended by MRP through an exception message suggesting planned order release and, therefore, conversion to a scheduled receipt. This recommendation will be made when the planned order release reaches the current time period (or action bucket). The release date is based on the offsetting of the planned lead time to the need date (planned order receipt) of the item. The MRP system does not consider capacity nor does it use safety stock in its recommendation.

Table 6-13. Conventional Plan Which Reacts to Gross Requirement Change

Parent A—Original Plan							
	Weeks						
	0	1	2	3	4	5	6
Gross requirement		15	15	15	15	15	15
Schedule receipt			30				
Projected on hand	30	15	30	15	0	-15	-30
Net requirements						15	15
Planned order receipt					30		30
Planned order release			30		30		
Projected available balance	30	15	30	15	30	15	30
Component B							
Gross requirement			30		30		
Scheduled receipt Projected on hand	40	40	10	10	−20		
Planned order receipt					100		
Planned order release		100					

Parent A—Week 5 Gross Requirement Increased to 25

	Weeks						
	0	1	2	3	4	5	6
Gross requirement		15	15	15	15	25	15
Scheduled requirement			30				
Projected on hand	30	15	30	15	0	−25	−40
Net requirement						25	15
Planned order receipt					30	30	
Planned order release			30	30			
Projected available balance	30	15	30	15	30	35	20
Component B							
Gross requirement			30	30			
Scheduled receipt							
Projected on hand	40	40	10	−20			
Planned order receipt				100			
Planned order release	100						

Table 6-14. Parent A—Week 5 Gross Requirement Increased to 25 with Weeks 4 and 6 Planned Order Receipts Firmed

	Weeks							
	0	1	2	3	4	5	6	7
Gross requirements		15	15	15	15	25	15	15
Schedule receipts			30					
Projected on hand	30	15	30	15	0	-25	-40	-55
Net requirements						25	15	15
Planned order receipt					30[a]		30[a]	30
Planned order release			30		30	30		
Projected available balance	30	15	30	15	30	5	20	35

[a]Firmed planned orders.

Purchase Orders

The conventional method of purchase order release has been a two-step operation in the past. The first step would be a planner reacting to the MRP release recommendation and generating a purchase requisition. The requisition would state the item, the required quantity, and the need date. The requisition would be forwarded to a buyer who would choose the supplier and create a purchase order. Because the planned lead time might allow for receiving, inspection, stocking, and, in some cases, safety time, the due date on the purchase order would not match the need date of the item. At the time of requisition release, that information must be fed to the system to convert the planned order to a scheduled receipt. When the purchase order is placed, the system must convert the system control from the requisition to the purchase order.

The two-step purchasing method is time-consuming and cumbersome. A more efficient approach being used more and more is the use of the buyer-planner. Purchased items that have repeat usage are negotiated for future requirements and controlled through a blanket purchase order. The blanket order is a long-term commitment with short-term releases to follow. The buyer-planner will react to the MRP release recommendation by dealing directly with the supplier and placing the short-term release in the form of a purchase order or a blanket release. Either form is related to the blanket purchase order. This approach saves order placement time, frees up the buyer for more meaningful activities, and reduces the computer input from two to one document.

Manufacturing Orders

The recommended release of manufacturing orders is based on the same due date—planned lead-time offset as is the purchase order release. Prior to actual release, it is desirable that the planner have an understanding of the current work-in-process situation; and if there might be any capacity constraints. Theoretically, the master schedule has been tested for reasonableness, but things do happen. Another check should be for material availability, especially for subassembly and assembly orders. Once the capacity and materials have been considered and the order seems doable, the planned order is converted into a scheduled receipt in the form of a shop order (work order). At the time of shop order generation, the materials required for the order are allocated in the inventory records. When the material is physically removed from stock, the material transaction must reduce both the on-hand quantity and the specific allocation. With the shop order release, related items such as drawings, bills of material, route sheets, move tickets, labor tickets, pick tickets, and so forth are generated. Which specific items to be included are dependent on the execution system.

The detailed scheduling of a manufacturing order is accomplished through the capacity requirements planning module. The planned lead time of an item is an estimate of total manufacturing time required, whereas the CRP detailed schedule calculation considers the lot size, the routing information (each work center's run and setup time), and predetermined queue and move estimates. The CRP calculated time may will not match the planned lead time, but it should be close. More important will be how close the actual lead time will be.

There are two methods of shop order scheduling—forward or backward. With backward scheduling, the calculation is initiated at the due date and back scheduled through the required operations. The start date may vary from the planned start date due to the above-mentioned difference between calculated and planned lead times. Backward scheduling is most applicable when scheduling components with differing lead times, but is required for assembly at the same predetermined time. Forward scheduling is initiated at the planned release date and forward scheduled through the operations. With forward scheduling, the completion date may differ from the planned due date. Forward scheduling is useful when bottleneck control is desired, especially if the bottleneck is the gateway operation or close to it.

With either method, if the difference in calculated and planned lead time is critical, lead time compression can correct the situation if capacity is not too strained. Most capacity requirements planning (CRP) softwares will allow the user to choose by item either forward or backward scheduling.

RESCHEDULING ORDERS

The MRP system will read when the due date of a scheduled receipt does not meet the need date of the requirement. This can be brought about by such things as master production schedule revisions, unplanned customer orders, engineering change orders, inventory adjustments, and/or scrap. When this happens, the MRP will supply an exception message suggesting rescheduling or canceling the order. Again, the MRP assumes infinite capacity and is informing the user of what changes are required to meet the existing plan. The user has four choices:

1. Reschedule the order as recommended.
2. Find another way to cover the requirement.
3. Change the existing plan.
4. Do nothing and hope that he or she will not be blamed when the plan is not met. (With this choice, it is recommended that the user's resume be up to date).

In order to make the proper decision, an analysis of the existing situation is required. Failure to move back or cancel an order will create excess inventory which, although undesirable, is not as bad as missed customer orders or assembly orders short parts. If the recommendation is to move up an order, the supplier or the shop should be consulted and commit to the change before the actual rescheduling.

If the recommended change is doable, the problem is solved. However, often times, the recommended change is impossible. An example would be an entire lot being scrapped and the MRP then calling for 1-week delivery of an item with a planned lead time of 12. In situations of this nature, the ability to understand all ramifications is paramount.

Interdependent Relationships

When an item is not going to meet a calculated need date, an understanding of that item's relationship to all affected material is required. If the item is an end item with independent demand, all lower-level components may be affected in that their previously required due dates may no longer be valid. Dependent demands may have either vertical and/or horizontal interdependencies. When a dependent demand item has missed or is predicted to miss the due date, one or both types of dependencies may come into plan.

Figure 6-1 is a simple illustration of vertical interdependency. Item A is dependent on the availability of item B, which, in turn, is dependent on the availability of item C.

Figure 6-2 is an illustration of both vertical and horizontal interdependencies. Item B's dependency on item C and item D's dependency on item E are both considered vertical interdependencies, whereas items B's and D's relationship to each other is a horizontal interdependency. When a material availability problem arises, analysis of all interdependencies is required.

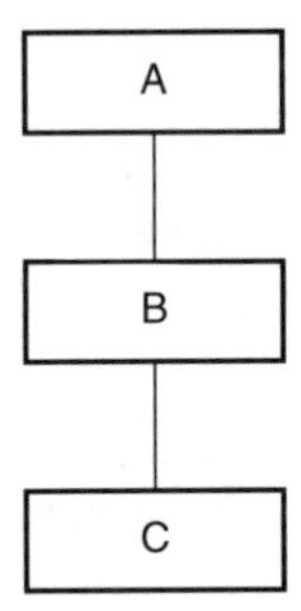

Figure 6-1. Vertical interdependency.

Bottom-Up Replanning

Solutions to availability problems may be increasing capacity, compressing lead time, material substitution, reducing order lot size, and safety stock or safety time usage. In order to make the best possible solution, all interdependencies should be reviewed. Bottom-up replanning is a process of analysis that starts at the lowest affected level of the bill of material. Through the use of pegging data, the effect on higher-level parents can be evaluated. This approach may also give information that will call for deexpediting, such as in Figure 6-2. If item B is not going to be available on the due date, item D's due date can be moved back to coincide with item B's revised due date. This is based on the assumption that item D has no other requirements. The process of exploding down to the lowest level and then bottom-up replanning is useful in what-if analysis.

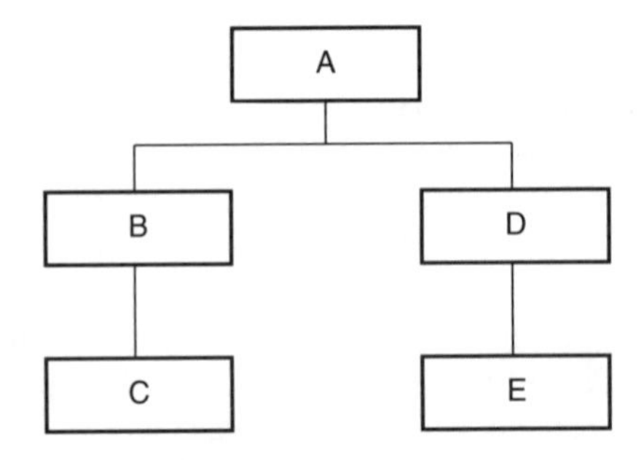

Figure 6-2. Vertical *and* horizontal interdependencies.

Table 6-15. Planned Order for Lot Size = 800, Lead Time = 2 Weeks, Safety Stock = 25

	Week							
	0	1	2	3	4	5	6	7
Gross requirement				500				600
Projected available balance	510	510	510	810	810	810	810	210
Net requirement								
Planned order receipt				800				
Planned order release		800						

CASE STUDY

Problems

1. The MRP has suggested releasing a planned order based on the information in Table 6-15. Would you release the order or is there an alternative?
2. A manufactured item with a planned lead time of 8 weeks has a planned order receipt on June 18 and a planned order release on April 23. On April 20, you release a shop order with a completion date of June 18 and with backward scheduling. The CRP system calculates a start date of March 17—over 4 weeks ago. What went wrong?
3. You have calculated a purchased part's lot size of 100 based on the EOQ formula using the annual usage, your carrying cost, and your ordering cost. The supplier of the item has agreed to accept the lot size of 100, but at a premium price. An order quantity of 1000 offers a substantial selling price reduction. What has caused the disparity in quantity costs?

Solutions

1. Accepting the MRP suggestion would cause the lot size of 800 to sit in stock 4 weeks before being used. The alternative is to use 10 of the 25 safety stock units and to stabilize the system by moving and firming the planned order receipt in week 7.
2. The planned lead time of 8 weeks is wrong. If it has been accurate in the past, there must have been a change in the process in the routing file which, in turn, is reflected in the detailed scheduling. A second possibility is that the queue and move estimates in the system have been increased due to a high level of work in progress. In any case, the lead times should be reconciled.

3. Your economic quantity is based on your cost of ordering—relatively low. Your supplier's costs are based on his setup cost. Because the purchased part is an extension of your operation, the manufacturing setup cost is a consideration. The textbook solution is to work with your supplier to reduce the setup cost.

QUIZ

1. A firm planned order "freezes" the
   I. quantity
   II. timing

   a. I
   b. II
   c. I and II
   d. Neither I nor II

2. If the setup time is decreased and the economic order quantity is recalculated, the quantity will
   a. increase
   b. decrease
   c. remain the same

3. The fixed order quantity technique will vary
   a. the order interval
   b. the quantity
   c. quantity and interval
   d. neither quantity nor interval

4. If you wish to minimize inventory, the lot size technique that will be most effective is lot for lot.
   a. True
   b. False

5. The EOQ calculation allows intermittent demand and does not require a steady demand rate.
   a. True
   b. False

6. Before rescheduling a manufacturing order, the planner should consider
   I. safety stock
   II. the parents requirements
   III. capacity

   a. I
   b. I and III
   c. II and III
   d. All of the above

7. The following factors affect the computation of requirements:
   I. Product structure
   II. Lot sizing
   III. Lead times
   IV. Timing of end item requirements

   a. I and II
   b. I and III
   c. I, II, and III
   d. All of the above

8. To determine the effect a component change has on a parent, you would use
   a. MRP
   b. capacity planning
   c. pegging
   d. shop-floor control

9. An allocated quantity in the inventory file indicates that the quantity is earmarked for
   a. a parent order
   b. a cycle count
   c. purging
   d. review

10. A planned order is the same as a scheduled receipt.
    a. True
    b. False

BIBLIOGRPAHY

APICS Dictionary, 7th ed. Falls Church, VA: American Production and Inventory Control Society, 1992.

Fogarty, D.W., Blackstone, J.H., Jr., and Hoffman, T.R., *Production and Inventory Management*. Cincinnati: South-Western Publishing, 1991.

Lunn, T. with Neff, S.A., *MRP—Integrating Material Requirements Planning and Modern Business,* Homewood, IL: Business One Irwin, 1992.

St. John, Ralph E., *Material and Capacity Requirements Planning Centrification Review Course,* Falls Church, VA: American Production and Inventory Control Society, 1991.

Vollman, J.E., Berry, N.L., and Wybark, D.C., *Manufacturing and Control Systems,* 3rd ed. Homewood, IL: Richard D. Irwin, 1992.

7
Capacity Planning

The successful execution of a material requirements plan (MRP) is dependent on a complete understanding of the required capacity for the plan. In the early days of MRP, the practitioner was happy to have the knowledge of valid component requirements and paid too little attention to the ability to supply those requirements. Capacity analysis is initiated at the strategic planning level and may conclude at the work center operation level. Figure 7-1 illustrates the hierarchy of capacity planning.

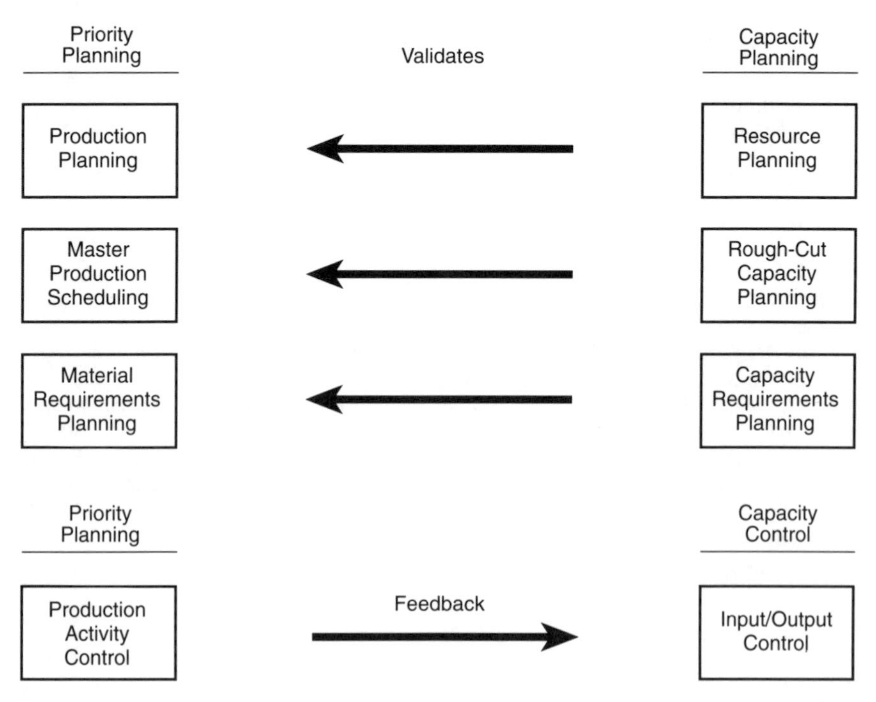

Figure 7-1. Capacity–priority interrelationships. [Reprinted with the permission of APICS Inc., *Capacity Management Review Course,* T. Bihun and J. Musolf, 1985.]

STRATEGIC PLANNING

The highest level of capacity planning is based on the long-range business plan that extends out as much as 5 years. This time period is necessary for planning cash requirements, additional plants, capital equipment, and, in some cases, a work force requiring specialized skills. The capacity management technique at this level is resource requirements planning and will be based on the analysis of total planned sales or broadly defined product groups. The projection of requirements at this level may be based on the proration and extension of critical resources. This proration technique is known as CPOF (Capacity Planning Using Overall Factors).

An example of long-range resource planning is a strapping machine manufacturer with a predicted annual growth rate of 15% for the next 5 years. The present annual sales level is 1700 machines and next year's forecast is 2000 machines (17% increase). The plant is presently operating at an estimated 60% of capacity. The 5-year projection of overall plant capacity is shown in Table 7-1. Resource planning indicates that additional overall plant capacity will be required within 3 years.

Table 7-1. Plant Capacity Projected

	Year					
	Present	1	2	3	4	5
Machine sales	1700	2000	2300	2650	3050	3500
Plant capacity %	60	70	81	93	107	123

The next level of capacity planning would be the analysis and validation of the intermediate 2-year production plan. Initial analysis would be based on product families or groups and the proration of critical resources. Using the strapping machine example, assume that the following resources are considered critical:

1. Cash for inventory investment
2. Work force
3. Turret lathes

The 2-year quarterly family forecast is shown in Table 7-2.The analysis of the intermediate forecast is accomplished by the utilization of a bill of resources—a listing of the key resources needed to manufacture one unit of a product family. The key resource bill for the strapping machine product groups is listed in Table 7-3.

Table 7-2. Two-Year Forecast

	Quarterly Forecast							
	1	2	3	4	5	6	7	8
Large steel strapping machine	50	50	50	50	50	50	50	50
Small steel strapping machine	235	250	250	265	265	270	280	285
Plastic strapping machines	115	125	125	135	135	155	195	215

Table 7-3. Key Resource Bill

Product Group	Inventory Investment Dollars	Labor Force Hours	Turret Lathe Machine Hours
Large steel strapping machine	10,000	175	2.0
Small steel strapping machine	6,000	125	2.5
Plastic strapping machine	3,000	100	6.0

Table 7-4. Product Group Explosion

	Quarters							
	1	2	3	4	5	6	7	8
Inventory investment in dollars (000)								
Large steel machines	500	500	500	500	500	500	500	500
Small steel machines	1,408	1,500	1,500	1,588	1,588	1,620	1,680	1,708
Plastic machines	344	376	376	404	404	464	584	644
TOTAL	2,252	2,376	2,376	2,492	2,492	2,584	2,764	2,852
Labor force in hours								
Large steel machines	8,750	8,750	8,750	8,750	8,750	8,750	8,750	8,750
Small steel machines	29,375	31,250	31,250	33,125	33,125	33,750	35,000	35,625
Plastic machines	11,500	12,500	12,500	13,500	13,500	15,500	19,500	21,500
TOTAL	41,745	52,500	52,500	55,375	55,375	58,000	63,250	65,875
Turret lathe machine hours								
Large steel machines	100	100	100	100	100	100	100	100
Small steel machines	588	625	625	663	663	675	700	712
Plastic machines	690	750	750	810	810	930	1,170	1,290
TOTAL	1,378	1,475	1,475	1,573	1,573	1,705	1,970	2,102

Explosion of the product group forecast based on the bill of resources determines the resource requirements as shown in Table 7-4. This intermediate resource planning projects the following:

1. Within the next 2 years, inventory investment will increase approximately $600,000.
2. Based on 500 hours per quarter per employee, the estimated work force requirement will increase from 84 to 132 employees in the next 2 years.
3. The existing turret lathe work center consists of two machines, each having an available capacity of 840 machine hours per quarter. By the sixth quarter, the work load will be greater than capacity and a third machine must be added to the work center or an alternate solution implemented such as outside sourcing or rerouting.

THE BILL OF RESOURCES

The bill of key resources utilized in the intermediate planning stage of the strapping machine analysis is one example of its use. The bill of resource is also used in rough-cut capacity planning. The structure and level of detail in the resource bill is based on the following:

1. What are the key resources requiring consideration? The user can structure any resource considered critical to the success of the plan.
2. Which level of planning is to be considered? For long-range or intermediate planning, the key resource bills will relate to broad product groupings or families, whereas the rough-cut planning level will require bills relating to the items in the master production schedule.

Both the CPOF (capacity planning using overall factors) and the basic bill of resources calculate the resource requirements in the same time period as the master scheduled family or item; that is, there is no time phasing. Consideration of time phasing is accomplished with the resource profile approach, a refinement to the bill of resources. Each key resource is time phased through the use of a resource profile.

The strapping machine bill of resources shown in Table 7-3 does not consider timing. If it is determined that the additional financing for inventory will be required two quarters before machine sales and that parts fabrication (the turret lathe operations) and the labor hours take place in the quarter prior to machine sales, the machine product group's resource profile will be as shown in Table 7-5.

Explosion of the strapping machine product group forecast utilizing the resource profile approach (time phasing the resource requirements) offsetting the resources is shown in Table 7-6 and reflects more timely resource requirements.

Table 7-5. Resource Profile

Resource	Quarters Before Sales Date		
	2	1	0
Inventory investment in dollars			
Large steel machines	10,000		
Small steel machines	6,000		
Plastic machines	3,000		
Labor force in hours			
Large steel machine		175	
Small steel machine		125	
Plastic machine		100	
Turret lathes in machine hours			
Large steel machines		2.0	
Small steel machines		2.5	
Plastic machines		6.0	

The resource planning based on the resource profile approach compared to the bill of resource explosion shows the following differences:

1. The inventory investment increase of $600,000 will take place in 1½ years rather than in 2 years.
2. The requirement for 132 employees will be in seven quarters rather than eight.
3. Additional turret lathe capacity will be required in the fifth quarter rather than the sixth.

When capacity planning at either the strategic long-range business plan or the intermediate production plan level indicates a resource increase, verification of the following factors is needed:

1. Is the increased level of activity permanent?
2. Is the capacity required unit of measure compatible with the capacity available unit of measure? If the bill of resources is expressed in standard machine hours, the available capacity cannot be expressed in scheduled hours.
3. Because the time periods are expressed in quarters or months, does the available capacity allow for demand variation within the time period? If a work center demand may vary by 20% on a week-to-week basis, a projected quarterly 90% utilization will not allow for proper schedule reliability.

Table 7-6. Time-Phased Product Group Explosion

	Quarters							
	1	2	3	4	5	6	7	8
Inventory investment in dollars (000)								
Large steel machines	500	500	500	500	500	500		
Small steel machines	1,500	1,588	1,588	1,620	1,680	1,708		
Plastic machines	376	404	404	464	584	644		
TOTAL	2,376	2,492	2,492	2,582	2,764	2,852		
Labor force in hours								
Large steel machines	8,750	8,750	8,750	8,750	8,750	8,750	8,750	
Small steel machines	31,250	31,250	33,125	33,125	33,750	35,000	35,625	
Plastic machines	12,500	12,500	13,500	13,500	15,500	19,500	21,500	
TOTAL	52,500	52,500	55,375	55,375	58,000	63,252	65,875	
Turret lathe machine hours								
Large steel machines	100	100	100	100	100	100	100	
Small steel machines	625	625	663	663	675	700	712	
Plastic machines	750	750	810	810	930	1170	1,290	
TOTAL	1,475	1,475	1,573	1,573	1,705	1,970	2,102	

ROUGH-CUT CAPACITY PLANNING

The term rough-cut capacity planning (RCCP) is sometimes used in the review process of the business or production plans, but in this book it is defined as the capacity technique to validate the master production schedule. In this context, rough-cut capacity planning requires the following:

1. The bills of resource must be based on master scheduled items or groups of very similar master production schedule (MPS) items.
2. If lot sizing is lot-for-lot, the resource profile time periods should be the same as the MPS—weekly.
3. If production lot sizes are large and not directly related to weekly requirements, the resource profile must be expressed in weeks or months and the MPS requirements grouped in similar time buckets. In this situation, the available capacity must be somewhat greater to allow for week-to-week demand variation.

Although rough-cut capacity planning at the MPS level gives more detailed information than resource planning at the production planning level, it does not project exact demand requirements because existing inventory is not considered, time buckets may be larger, and time phasing is based on less

Table 7-7. Rough-Cut Capacity Planning Explosion

	MPS in Weeks											
	1	2	3	4	5	6	7	8	9	10	11	12
Large steel machines		10			10			10			10	
Small steel machines	20	20	20	20	40	20	20	20	20	20	20	20
Plastic machines	10	10	10	10	10	10	20	10	10	10	10	10

	MPS in Months		
	July	August	September
Large steel machines	10	20	10
Small steel machines	80	100	80
Plastic machines	40	50	40

	Turret Lathe Machine Hours in Months		
	May	June	July
Large steel machines (2 machine h ea.)	20	40	20
Small steel machines (2.5 machine hrs. ea.)	200	250	200
Plastic machines (6 machine hrs. ea.)	240	300	240
TOTAL	460	590	460

detailed estimates than used in MRP. The following is an example of MPS validation through rough-cut planning compared to the resource profile of the production plan.

The production plan profile for second quarter turret lathe demand is projected to be 1475 machine hours (see Table 7-6). This was based on a third quarter sales forecast of 425 machines (see Table 7-3).

The master schedule for the third quarter is stated in 12 weekly time buckets, but for purposes of rough-cut capacity, planning is grouped in 4-week monthly buckets. The turret lathe time phasing used in the production plan profile was one quarter, whereas the RCCP was based on 2-month time phasing. The RCCP results are shown in Table 7-7.

The available capacity of the turret lathe work center was estimated to be 1680 machine hours per quarter compared to the production plan demand of 1475. However, the more detailed RCCP indicates a problem in the month of June with a demand of 590 hours compared to an available capacity of 560 (1680 ÷ 3). Consideration must be given to either adjusting the MPS or finding additional capacity, such as overtime.

CAPACITY REQUIREMENTS PLANNING

Capacity requirements planning (CRP) is the planning and control of the resources needed to produce the requirements generated by the MRP system. Detailed available capacity must be determined and compared to the anticipated capacity required. The process involves each work center and covers the same horizon and time periods as the MRP. Both open (released) shop orders as well as planned orders are considered in the determination of the required capacity (the work load). Detailed schedules for all orders are calculated based on the routing files, lot sizes, and work center files. These schedules, released and anticipated, are required in order to place each work center's load in the proper time period. As rough-cut capacity planning was used to validate the master production schedule, capacity requirements planning will validate the material requirements planning's output.

Available Capacity

Capacity is expressed as the capability of a worker, machine, or work center to produce output over a specific time period. The normal unit of measure is standard hours. The available capacity formula is

$$\text{Capacity available} = \text{Hours available} \times \text{Utilization} \times \text{Efficiency}$$

The hours available are the hours scheduled for work. A three-machine work center scheduled for two 8-hour shifts 5 days per week has 240 ($3 \times 2 \times 8 \times 5$) hours available for work.

Machine utilization is an estimate of the percent of the available scheduled hours the machine actually will be running. The standard formula for utilization is actual hours run/hours available. The downtime of a machine can be due to machine breakdown, absenteeism, or lack of work or material. Absenteeism and machine breakdowns are negatives and the goal should be to eliminate them completely. Lack of work in some situations may be planned to allow for the random arrival of work from other work centers. It is not realistic to set a goal of 100% utilization of a machine or work center unless it is a bottleneck operation. In an effort to achieve high utilization, queues (work-in-process) have often been carried at unreasonably high levels. Unless the manufacturing environment is continuous flow, it is not reasonable to expect or plan for a completely balanced plant. John H. Blackstone, Jr., in *Capacity Management,* splits utilization into two measurements, availability and activation. Availability considers absenteeism and machine breakdowns with a goal of no lost time (factor of 1). The activation factor, a measurement of downtime due to lack of work, will be less than 1 except for a bottleneck. The activation factor of a bottleneck(s) should be 1, whereas the activation factor of a nonbottleneck should be only what is required to keep bottlenecks working at all times.

The efficiency factor measures the anticipated performance of the machine compared to the standard set for it. This factor is dependent on the skill of the operator as well as the condition of the machine, the material, tooling, and so on. The formula for efficiency is standard hours produced/actual hours charged. If the work center previously mentioned with a weekly hours available of 240 was rated with an efficiency factor of 110% and a utilization factor of 80%, the weekly rated capacity would be

$$240 \times 1.1 \times 0.8 = 211 \text{ hours}$$

Capacity Data Files

Capacity requirements are based on released and planned orders generated by the MRP system. Work-load calculations are based on relating these orders to (1) the routing file which details the required manufacturing operations and (2) the work center master files which detail the available capacity and lead-time (queue) allowances.

Each manufactured part has a separate routing by sequence of operations. The basic data are the operation number, the operation description, the planned work center, standard setup time, run time per unit, and tooling requirements. Alternate routings are also maintained by many manufacturers.

Each operation is loaded to a work center which has basic control factors listed in the work center files. The detailed data are work center identification and description, number of shifts scheduled, number of machines, hours per shift, utilization factor, efficiency factor, and queue time allowance.

Work-Load Calculation

The required capacity (work load) calculation considers both open and planned orders. The open (released) order information is in the open order detail file and lists by part number and work order, completed operations, and operations still open. Each remaining operation has been scheduled. The planned orders are scheduled by operation using either backward or forward scheduling as described in Chapter 6. The lead time for each operation is based on the lot size, the run time per unit, the setup time, the work center queue allowance, and wait and move allowances. The wait and move allowances can be maintained in either the work center master or the routing file. The lead-time data are required in order to place the work load in the proper time period. The work load in the time period is only the setup and run calculations as these

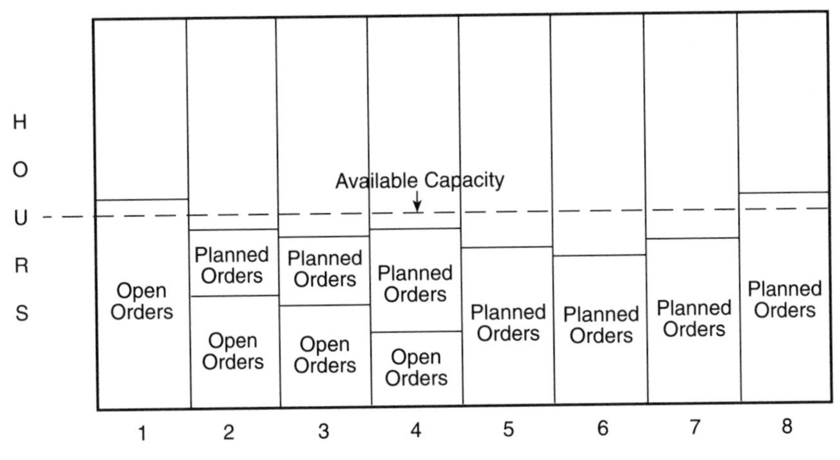

Figure 7-2. Work Center Load Profile

elements are the ones that use work center capacity. The queue, wait, and move elements do not use capacity; they just use up space and time.

The work center profile is developed by listing both open order and planned order loads by work centers and time periods and then comparing to available capacity. Because MRP assumes infinite capacity, overloads and underloads can be expected. If the detailed work center profiles are reasonable (i.e., doable), the MRP output is validated and the process of "closed-loop MRP" is complete. Figure 7-2 is an example of a work center profile.

Assume a 100-piece order has a setup time of 2.0 hours and a standard run time of 0.1 hour/piece and is scheduled to run in a work center that calls for a 2-day (8 hours/day) queue allowance, a 4-hour wait time, and a 4-hour move time. The work load in the work center would be $100 \times 0.1 = 10$ hours run + 2 hours setup = 12 hours. The scheduled lead time for the operation would be 16 hours queue + 2 hours setup + 10 hours run + 4 hours wait + 4 hours move = 36 hours.

CAPACITY MANAGEMENT

Lead Time and Queue Control

The lead time of each operation is the sum of queue (time waiting at work center), setup (time preparing for the operation), run (time performing operation), wait (time waiting to be moved on), and move (time moving to the next operation). The setup and run times are the elements using capacity; although they are the value-added functions, they are normally a small part

Table 7-8. Routing Example

Operation	Work Center	Setup Time in Hours	Run Hours	Queue Allowance Days
010	100	0.5	0.5	2
020	200	1.0	3.0	3
030	200	0.5	2.5	3
040	300	0.5	1.5	2
050	400	0.5	1.5	2
060	500	0.5	3.5	3

(5–20 percent) of the total lead time. Table 7-8 is an example of a typical six-operation routing.

Assuming a wait and move time of one-half day per operation, the total lead time in days is

$$3\text{ (wait and move)} + 2\text{ (setup and run)} + 15\text{ (queue)} = 20$$

The setup and run time is 10% of the lead time, whereas the queue time is 75% of the lead time. In the planning calculation, the queue time is an estimate, but it is the key to lead-time control.

The queue allowance for each work center is based on the anticipated number of jobs waiting to be processed. A queue of orders is needed to allow for the random variation of arrival of other orders. If a high machine utilization is the goal, there must be a large queue time allowance so as to ensure job availability at all times. Reducing the queue will reduce lead time and work in process and increase flexibility. Assuming random variation of arrivals and lot sizes, the only way to reduce the queue is to have capacity available for all arrivals which, in turn, calls for reduced machine utilization. The compromise between planned queue time and capacity utilization can be controlled through Blackstone's activation factor.

Capacity Problems

The queue allowances in the scheduling of shop orders are based on the assumption of a certain level of work-in-process. Figure 7-3 graphically indicates if the input is greater than the output, the plan (and plant) is overloaded, thus increasing both work-in-process and lead time. If the average work-in-process increases from 7 to 10 weeks, the average lead time will increase from 7 to 10 weeks. The rule is not to overload the plant but to keep the orders in production control rather than releasing and overloading.

If there are only one or two work centers, the orders can remain in the

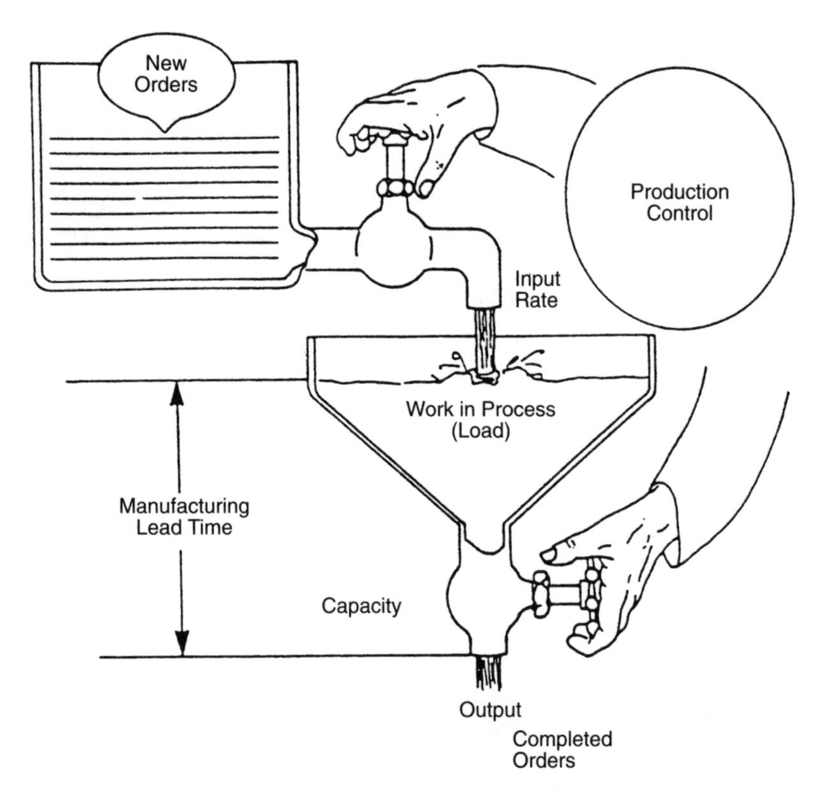

Figure 7-3. Capacity and work load. [Reprinted with the permission of APICS Inc., *Shop Floor Control Training Aid,* M. Hablewitz, 1979.

office, but with multiple work centers, work-in-process control can be most difficult. The plant may be in an overloaded condition due to temporary bottlenecks that affect only certain products or processes. If this is the situation, products not affected by the bottlenecks should not be held, but identification of the unaffected items can be difficult and time-consuming.

Products routed with multiple operations have a greater chance of missing planned operation dates for the operations at the end of the routing (the more operations previously scheduled, the more chances for things to go wrong). Figure 7-4 depicts a profile for a work center that performs operations at the end of an 8-week process. Week 6 through week 8 work load projections are based on planned orders and are shown to be in line with available capacity. In week 1, 750 hours were released for week 8 rather than the 720 planned hours. Going into week 8, the 750 hours scheduled for that week arrives on time, but 30 of the 710 scheduled for week 6 and 90 of the 680

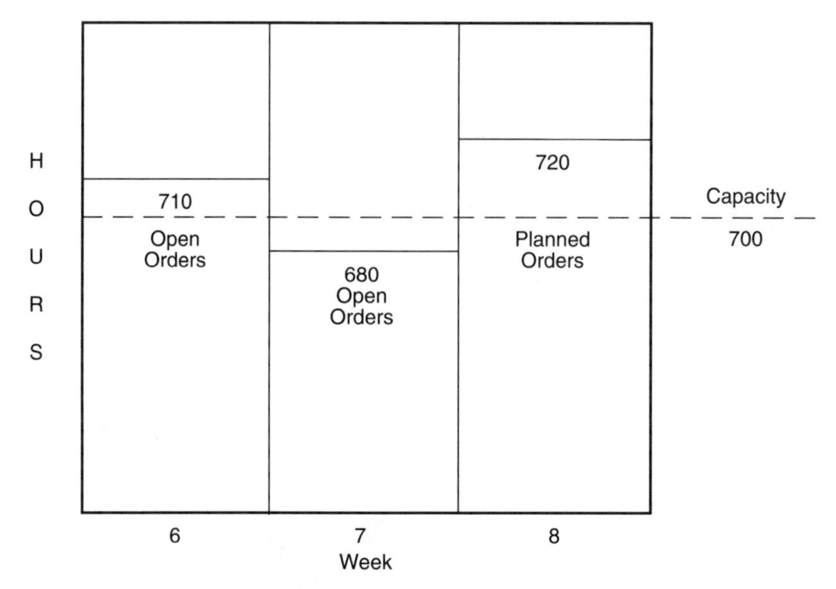

Figure 7-4. Work center profile.

scheduled for week 7 are late arrivals (past due). The required hours for week 8 are now 870 compared to a capacity of 700.

Monitoring of day-to-day activities is required in order to best manage work in process and lead times.

Input–Output Control

The actual input of work to a work center is compared to the planned input projected by the CRP system. The actual output is compared to planned output. The input difference will be due to the unplanned variations in arrival patterns due to previous operations being completed either early or late against schedule. Output less than planned can be due to production problems such as machine breakdowns, quality considerations, absenteeism, and so forth, or lack of available work. Output greater than planned is usually due to an effort to reduce the queue of available work.

The goal of input–output monitoring is to control a queue of work-in-process consistent with the capacity plan. An overloaded work center will cause late delivery and increased work in process, whereas a starved work center will create manufacturing inefficiencies. This measurement system is necessary to identify the previously described capacity problems.

Input–output data can be maintained on a daily or weekly basis. Table 7-9 is an example of a 4-week input/output record stated in standard hours.

Table 7-9. Input/Output Control

	Week			
	1	2	3	4
Planned input	110	115	135	130
Actual input	120	120	130	140
Cumulative deviation	+10	+15	+10	+20
Planned output	115	115	130	135
Actual output	110	120	125	130
Cumulative deviation	−5	0	−5	−10
Actual backlog	70	70	75	85
	Desired Backlog = 50			

The backlog going into week 1 was 60 standard hours. Analysis of the four-week data shows the following:

1. Planned input was 490 hours and planned output was 495 hours. If the plan had been met, the backlog would have been reduced from 60 to 55 hours.
2. Actual input was 20 hours more than planned, whereas output was 10 less than planned. Therefore, rather than reducing the backlog by 5 hours, it was increased 25 hours (60 to 85).
3. The queue allowance consistent with desired backlog is approximately 2 days $(110 + 115 + 135 + 130) \div 20$ days $= 24.5$/day; $50 \div 24.5 = 2 +$ days). The actual queue is over 3 days. Average queue $= 75$; $75 \div 24.5 = 3 +$ days.
4. Managerial action is required.

Scheduling

The releasing of work to the shop in the form of work orders will list operation control dates required to meet the demands of the MRP system. Whereas capacity work-load analysis has indicated that the plan is reasonable, the resultant schedule is still based on infinite capacity in that the accumulated load at each work center is not compared to the capacity of that work center. The success of infinite scheduling is dependent on the initial capacity management, the flow of orders, and shop-floor prioritizing at each work center.

Finite scheduling is a technique that will not allow work to be loaded beyond the stated capacity of a work center. The prioritizing of work is based on predetermined rules such as work center completion date or next operation availability. Infinite and finite scheduling concepts will be covered in detail in the execution chapters of this book.

CASE STUDY

Problems

1. A manufacturer of a single product estimates a growth rate of 7% per quarter for the next year. The production rate in the present quarter is 500 machines. There are two resource concerns:
   a. Trained Personnel. There are at present 102 trained operators as well as 10 trainees. The anticipated production rate is five machines per operator and the training time is 6 months. Should any trainees be added and, if so, how many?
   b. The Inventory Investment. The inventory investment per machine is $2000 and it takes 3 months to arrange financing. How much inventory investment should be planned for the immediate future?
2. The CRP work-load profile for a gateway (first operation) work center is as follows:

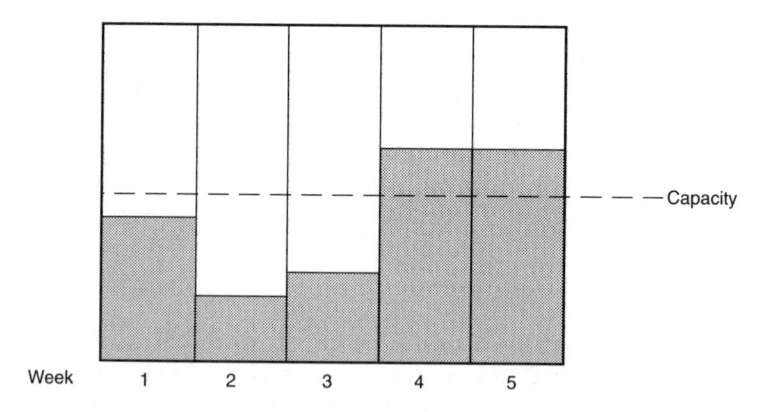

What actions would you recommend be taken by the planner including additional data analysis?

Solutions

1. Determine a key bill of resource for each machine.

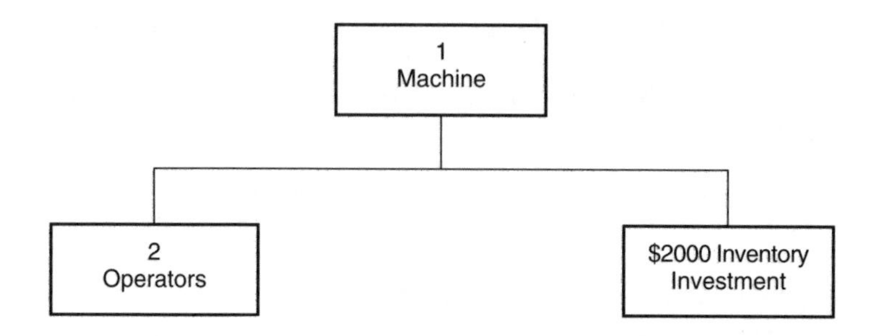

Project a master schedule based on 7% per quarter growth and explode using the bill of resource.

	Quarter				
	Present	1	2	3	4
Machines	500	535	572	612	655
Manpower	100	107	114	122	131
Investment in dollars (000)	1000	1070	1145	1225	1310

a. One hundred twenty-two (122) trained operators will be required in the third quarter. There are at present 112 operators and trainees. Ten more trainees should be hired.

b. An inventory investment of $1,145,000 will be needed in the second quarter and should be arranged for now.

2. Work scheduled to start in weeks 4 and 5 should be transferred to weeks 2 and 3 so as to level the work load and eliminate the overload. Before transferring specific jobs, it should be ascertained that the required raw material will be available at the revised date.

QUIZ

1. MRP systems assume that capacity considerations are in
   I. the master schedule
   II. the inventory files
   III. MRP logic
   IV. the bills of material

   a. I
   b. II
   c. I and III
   d. II and IV

2. A Material requirements planning system is designed to answer
   I. what can be produced with a given capacity
   II. what production is required to meet a given master production schedule

   a. I
   b. II
   c. I and II
   d. None of the above

3. Rough-cut capacity planning considers
   I. every work center
   II. key resources
   III. financial projections only

   a. I
   b. II
   c. I and II
   d. All of the above

4. The resource bill of material structure is based on
   a. single-level parts bill only
   b. multilevel parts bill only
   c. purchased parts
   d. whatever you want

5. A proper load projection has the following:
   I. Completeness
   II. Based on valid priorities
   III. Future visibility
   IV. Includes planned and open orders

   a. I, II, and III
   b. I, II, and IV
   c. I, III, and IV
   d. All of the above

6. The capacity of a work center is based on
   I. hours available
   II. machine utilization
   III. open orders
   IV. machine efficiency

   a. I and II
   b. I and IV
   c. I, II, and IV
   d. All of the above

7. The following are considered part of the work load:
   I. Queue time
   II. Run time
   III. Setup time
   IV. Wait time

a. I and II
b. II and III
c. I and III
d. All of the above

8. Random variation of work arrival at a work center can cause
   I. a build up of queue
   II. machine down time due to lack of work

   a. I
   b. II
   c. I or II
   d. Neither I nor II

9. In the lead-time calculation, random variation of work is compensated for by
   I. increasing the run estimate
   II. queue time allowance
   III. reducing the efficiency factor

   a. I
   b. II
   c. III
   d. All of the above

10. The activities of a work center are monitored through
    a. MRP
    b. the master schedule
    c. rough-cut capacity planning
    d. input–output control

BIBLIOGRAPHY

APICS Dictionary, 7th ed., Falls Church, VA: American Production and Inventory Control Society, 1992.

Bihun, T. and Musolf, J. *Capacity Management Review Course,* 1985.

Blackstone, J.H., Jr., *Capacity Management.* Cincinnati: South-Western Publishing, 1989.

Hablewitz, M. *Shop Floor Control Training Aid,* 1979.

St. John, Ralph E., *Material and Capacity Requirements Planning Certification Review Course,* Falls Church, VA, American Production and Inventory Control Society, 1991.

Vollman, J. E., Berry, N. L., and Wybark, D. C., *Manufacturing and Control Systems,* 3rd ed. Homewood, IL: Richard D. Irwin, 1992.

8
Distribution Resource Planning

A materials logistics system is defined as the planning and coordination of the physical movement aspects of operations from raw material to customer receipt with the goal of minimizing costs for the desired service level. If a manufacturing firm stocks finished goods in branch warehouses, the control of distribution requirements should be coordinated with the manufacturing system. This is accomplished with distribution resource planning (DRP). As with MRP and MRP II, there has been some confusion between the terms "distribution requirements planning" and "distribution resource planning." Distribution requirements planning refers to the time phased, net requirements explosion of warehouse needs via material requirements planning (MRP) logic. Distribution resource planning is an extension of distribution requirements planning and considers the resources of the distribution system in a manner similar to the capacity considerations covered in material resource planning (MRP II).

The warehouse stocked with finished goods to serve the customer is considered a distribution center (DC) and is synonymous with the term "branch warehouse." A warehouse serving one or more distribution warehouses is considered a regional distribution center (RDC). A regional distribution center may ship directly to customers as well as serving distribution warehouses. Although it might be desirable to serve west coast customers with local distribution centers, such as in San Francisco, Los Angeles, Portland, and Seattle, the shipping costs from a manufacturing center in Chicago to each center might be prohibitive. Establishing one of the locations as a regional distribution center to supply the other three (and to perhaps carry safety stock) might reduce the total distribution cost and justify the distribution centers. A regional distribution center is also known as a "feeder warehouse."

The warehouse supplying the distribution system is considered the central supply center and is usually at the manufacturing site. Like regional distribution centers, the central supply center may ship directly to customers as well as supplying distribution centers. Application of MRP principles and techniques call for the distribution center to be considered the parent with independent demand, whereas the unit supplying the DC (the regional distribution center or the central supply center) is similar to a component in that the demand is dependent upon the requirements of the parent.

CONCEPTS AND LOGIC

Reorder Point Limitations

Lacking a DRP system, many distribution operations are based on individual reorder point systems at each distribution center. This approach assumes product availability at the central supply center and is not integrated with the manufacturing system. If the demand of a single stocking location calls for an even, continuous rate, the reorder point system will work. However, if there are multiple stocking locations, each having even, continuous demand, the combined results will be lumpy requirements for the supplying facility.

Table 8-1 illustrates the time-phased combined demand of the distribution centers, each with even flow demand. The controlling factors at the DC centers are as follows:

Table 8-1. Combined Demand

Week	DC #1	DC #2	DC #3	Total
1	500			500
2		600		600
3			600	600
4				—
5		600		600
6	500			500
7			600	600
8		600		600
9				—
10				—
11	500	600	600	1700

	Weekly Demand	Order Quantity	Order Frequency
DC #1	100	500	5 weeks
DC #2	200	600	3 weeks
DC #3	150	600	4 weeks

Assuming that DC #1 requires shipment in week 1, DC #2 in week 2, and DC #3 in week 3, the combined central supply replacement requirements will be as shown in the total column of Table 8-1.

Without the visibility of DRP, the central supply center in the above situation would not realize what was going to happen in weeks 9, 10, and 11. With visibility of future demands, proper steps could be taken at the master planning level of the manufacturing system.

Time-Phased Requirements

Due to the intermittent or lumpy demand pattern at the central supply center, the system must communicate and combine the individual demands of each distribution center in a time-phased mode. The forecasted independent demand of each DC may be continuous or discontinuous, but, in either case, the forecast is treated as a time-phased order point gross requirement similar to the gross requirement control of the master production schedule. As with MPS gross requirements, there will be the decision process required as how to treat customer orders in a forecasting environment. As with the master production schedule (MPS) process, the control will be dependent on time-fence policies and order quantity comparison to forecasted quantities.

Once the gross requirements are established at the DC level, the net requirements and planned orders are calculated using MRP time-phasing logic. The planned order release of the DC parent is then exploded down to the next level which will be either a regional distribution center or the central supply center. The level-to-level explosion, the parent-to-component relationship, and the dependent demand concepts are exactly the same with DRP as with MRP. However, because MRP deals with a manufacturing environment and DRP in a distribution environment, the data elements in the record files go by different names but react to the same logic. Table 8-2 lists the MRP elements and their equivalent DRP elements.

Table 8-2. Data Element Comparisons

MRP	DRP
Gross requirement	Forecast
Scheduled receipt	In transit
Projected available balance	Projected available balance
Planned order receipt	Planned shipment receipt
Planned order release	Planned shipment release

Safety Stock

Safety stock in a distribution system may be an educated guess based on anticipated reliability (or unreliability) of both forecast and product on-time delivery. A more sophisticated approach is to measure forecast and supply error and relate it to desired customer service level (see Error Measurements in Chapter 4). A decision must be made as to the location of the safety stock. If the safety stock is located at the central supply center, less stock is required, as there is less uncertainty of demand at the center where positive and negative errors are combined. However, the response to the customer is not as immediate and the cost of air freight may negate the savings in inventory investment.

Safety stock may be controlled through a specific quantity expressed in units or weeks of forecasted demand. A second technique is to maintain a safety time in the calculation. Safety quantity is controlled by not letting the projected available balance fall to less than the safety quantity rather than to zero. Safety time is controlled by moving the planned shipment receipt date forward by the safety time factor. If the demand is continuous and even, the results of either system is the same. However, safety time is favored when

1. there is intermittent usage and average usages mean nothing or
2. when usage rates change quite often and quantity recalculations are required.

As with manufacturing systems, the use of safety stock requires understanding and caution. The user must not only know when to utilize the safety stock but how to control its replenishment. When there is an unexpected increase of business throughout the distribution network, the manufacturing facility may be struggling to increase the output to the new level of demand, let alone trying to replace the safety stock. In this situation, the safety stock numbers must be adjusted to allow for controlled replacement.

TIME-PHASED PLANNING

Production Planning

The production plan of a manufacturing facility (the central supply center) is based on product groups or families in the distribution system. The plan will cover the distribution demands of the product groups for 1 to 2 years as well as adjusting for anticipated inventory changes. The planned inventory changes may be to allow for seasonality, plant vacation shutdowns, inventory reduction, and so forth. Once established, the production plan is validated by determining the requirements of the key resources needed to produce the plan. The grouping of product families is based on end items that are similar in design, requiring similar manufacturing facilities, and whose demands can be monitored as a group.

Master Scheduling

The master production schedule is by end item and covers all manufactured end items in the distribution system. The gross requirements of master scheduled items will equal the sum of the planned shipment releases of the distribution systems as well as additional local demands at the central supply center.

DRP

DC 1 Product A LT = 2

Weeks	1	2	3	4	5	6
Forecast	20	20	20	20	20	20
In Transit		40				
Projected Available Balance	0	20	0	20	0	20
Planned Shipment Receipt				40		40
Planned Shipment Release		40		40		

DC 2 Product A LT = 2

Weeks	1	2	3	4	5	6
Forecast	30	30	30	30	30	30
In Transit	60					
Projected Available Balance	30	0	40	0	40	0
Planned Shipment Receipt			70		80	
Planned Shipment Release	70		80			

MPS (Central Supply) Product A LT = 0

Weeks	1	2	3	4
Gross Rrquirements	70	40	80	40
Scheduled Receipts	80			
Projected Available Balance	10	50	50	10
Planned Order Receipt		80	80	
Planned Order Release		80	80	

MRP

Component B LT = 1

Weeks	1	2	3	4
Forecast		80	80	
In Transit				
Projected Available Balance	0	0	0	0
Planned Shipment Receipt		80	80	
Planned Shipment Release	80	80		

Component C LT = 2

Weeks	1	2	3	4
Forecast		80	80	
In Transit		100		
Projected Available Balance	0	20	40	40
Planned Shipment Receipt			100	
Planned Shipment Release	100			

Figure 8-1. Integrating DRP and MRP.

For purposes of this analysis, it is assumed that the combined distribution demand is the only demand. The control of the MPS is as explained in Chapter 4—The Master Production Schedule.

While in an MPR system, the MPS end item demand is defined as independent, the same MPS demand in a DRP system is dependent on the demands of the distribution centers. The MPS is the link or interface between DRP and MRP.

Integrated Distribution and Manufacturing Systems

Although the MRP system is driven by the master production schedule, that same master production schedule is driven by the needs of the distribution system. Figure 8-1 is an example of the integrated systems where two DCs require product A, which is assembled with components B and C.

If the distribution centers have separate data bases, a program will have to be written to read the planned shipment release requirements from all the centers, sum them, and post them to the MPS. Some software systems have DRP systems which can be fed to the MPS/MRP system.

THE DISTRIBUTION CENTER

Inventory Controls

Each part number stocked at a distribution center will be controlled by planning control values pertinent to the specific distribution center. The lead time is the sum of the supplying center's order release, pick, and load time, in-transit time, and the DC unloading and stocking time. The planned lot size of each part is based on the following considerations:

1. Economic ordering. The traditional balancing of the costs of ordering and the costs of carrying the inventory.
2. Frequency of shipments to the distribution center. The individual item's needs must be combined with other items in order to allow for economical truck or rail shipments.
3. The pallet size of the part.
4. Weight and cube to assist is best use of truck or rail capacity.

If the policy is to carry safety at the DC, the quantity or safety time control numbers should be based on the same criteria as explained under Error Measurement in Chapter 4, but relative to the variations at the individual DC. The usage rate at the DC (the gross requirements) will be the time-phased forecast of DC demand adjusted up close for open orders. In some DRP systems, the part number at a DC will be the corporate part number plus a suffix unique to the DC. This allows the combining of requirements at the supply center's master schedule level, but at the same time, the identification of individual requirements for each DC.

The Planning Horizon

The planned shipment release is determined by offsetting the planned shipment receipt by the lead time of the DC. The DC lead-time increment is an added time value to the critical path required for manufacturing the product. Therefore, the DRP planning horizon should be the sum of the MPS/MRP planning horizon and the DC lead time. An example of this concept is an item with a manufacturing critical path of 26 weeks and distributed overseas with a transportation time of 13 weeks. An overseas requirement at the end of September must be in stock at the central supply center by the end of June. To produce the item by the end of June, the initial action (probably placing a purchase order) would have to be placed at the beginning of the year. The 39-week DRP horizon is necessary to predict the end of June stock requirement to the MPS/MRP system, which, in turn, must use its 26-week horizon to predict the beginning of the purchasing requirements. Due to lead times differing with transport modes, the planning horizon must be flexible so as to accommodate changing values.

Action Messages

The DRP system will generate action messages to the DC to release shipment orders to its demand source (the regional or central supply center) when the release data coincides with the action bucket. It will also call for action when an in-transit order is scheduled too early or too late for requirements. The logic is the same as the action messages in MRP relating to scheduled orders. The difference is that a scheduled order may be more easily expedited or deexpedited, whereas an in-transit order's time is quite fixed. The DRP in-transit action message is therefore more useful as an information tool.

THE CENTRAL SUPPLY CENTER

The Master Production Schedule

Whereas the MPS of the central supply center drives the MRP, it, in turn, can be considered driven by the DRP system. The goal of the MPS is to create a plan that will do the following:

1. Meet the demands of the distribution centers
2. Be compatible with manufacturing capacity
3. Call for a level manufacturing output
4. Allow for planned inventory reductions or buildups
5. Be consistent with planned shipping schedules

Pegging and Firm Planned Shipments

The master scheduler will use pegging routines with the DRP system to trace back detailed demand sources. This will be useful when adjustments to the MPS are called for in order to meet MPS goals. The routine of pegging up through the DRP levels is similar to pegging up through the MRP levels when tracing the source of a manufacturing demand.

The use of firming planned shipments is similar to firming planned orders in an MRP system in that this routine is for overriding the planning logic for DRP. The master scheduler may wish to override the system when using the MPS to allow for planned buildups, temporary lot size reductions, planning to use safety stocks to meet demands, or adjusting demands to coincide with shipping schedules. The firming of planned shipments requires close communication between the central supply center and the affected distribution center.

Action Messages

Whereas the DRP system will generate action messages to the DC to release shipments to the central supply center, the MPS system will generate similar action messages to release MPS orders. The MPS system will also recommend changing MPS order due dates even if they are firmed. Although it may be impractical if not impossible to change the MPS, it communicates information that distribution needs are in trouble and some type of action is required.

Solution Examples

Using the DRP system for stock buildups calls for a complete understanding of the distribution network. The purpose of the buildup will be to solve distribution problems such as uneven demands due to seasonality, promotions, or special sales, changing distribution networks, concern over labor unrest at the manufacturing center, or planned plant shutdowns. The stock buildups can be at the central supply center, the distribution center(s), or both. The buildup considerations can be as follows:

1. Space availability
2. The capacity of the DC to handle the work load
3. Transportation availability
4. Inventory-carrying costs

Table 8-3. Temporary Stockpile Control

Original Master Schedule Display Safety Stock = 0							
	Month						
	1	2	3	4	5	6	7
Distribution demands	2000	2000	2000	2000	2000	2000	2000
MPS receipt	3000	3000	3000	3000	—	—	2000
Projected available balance	1000	2000	3000	4000	2000	—	—
Stockpile Location DRP Display							
Forecast					2000	2000	
Projected available balance	1000	2000	3000	4000	2000	—	—
Firm planned shipment	1000	1000	1000	1000			
Adjusted Master Schedule Display							
Distribution demands	3000	3000	3000	3000	—	—	2000
MPS receipt	3000	3000	3000	3000	—	—	2000
Projected available balance	—	—	—	—	—	—	—

An example of establishing a stockpile location (a temporary DC) in order to handle a 2-month central supply buildup due to a plant renovation is shown in Table 8-3. The original master schedule reflects distribution demand of 2000 units per month for 7 months. The first 6 months' demand (12,000 units) are scheduled for production in the first 4 months so as to allow for the fifth and sixth month renovation shutdown. The central supply center lacks storage capacity so the stockpile location is established to accept same-day delivery from central supply at the rate of 1000/month. The stockpile location display reflects the 4-month buildup using firm planned shipments. It also shows the fifth and sixth month gross requirements when the location will act as a regional distribution center supplying the needs of the distribution system. With the addition of the stockpile location DC, the DRP demands will generate the adjusted MPS display shown.

A second example is when an existing distribution center is used to stock a plant buildup but cannot receive the buildup at the rate the buildup is being produced. Table 8-4 shows the MPS planned buildup where weeks 1, 2, and

Table 8-4. Vacation Buildup

Original Master Schedule Display
Safety Stock = 1000

	Week 1	2	3	4	5
Distribution demands	1000	1000	1000	1000	1000
MPS receipt	1500	1500	1500	—	500
Projected available balance	1500	2000	2500	1500	1000

Distribution Center A Display
Lot Size = 400, Lead Time = 2 Weeks, Safety Stock = 100

Forecast	200	200	200	200	200	200	200
Projected available balance	300	100	300	100	300	100	300
Planned shipping receipt			400		400		400
Planned shipping release	400		400		400		

Table 8-5. Firming Buildup Shipments

Distribution Center A (After Shipment Adjustments)

	Week 1	2	3	4	5	6	7
Forecast	200	200	200	200	200	200	200
Projected available balance	300	100	500	300	700	500	300
Planned shipping receipt			600		600		
Firm planned shipments	600		600				

Adjusted Master Schedule Display

Distribution demands	1200	1000	1200	1000	600
MPS receipt	1500	1500	1500	—	500
Projected available balance	1300	1800	2100	1100	1000

3 produce an extra 500 units each week to allow for the week 4 vacation shutdown as well as reduced start-up production in week 5. Distribution center A's DRP plan is also shown.

The maximum space the plant (central supply) can handle is 2100 units. The decision is made to move 400 units of the build up to DC A. DC A cannot receive more than 600 units per week. The problem is solved by increasing and firming the planned shipments in weeks 1 and 3 to 600 units. Table 8-5 shows the results of this action.

TRANSPORTATION PLANNING

Shipping Requirements

Distribution resource planning (DRP) reflects the inventory needs of the distribution centers just as MRP reflects the material needs in manufacturing. As MRP must be compatible with available capacity, the DRP must not only be compatible with resources but also must satisfy transportation criteria so as to effect optimum shipping costs. For both truck and rail shipments, there are rate to weight structures with the lowest cost per pound achieved through full rail or truck loads. The goal is to maximize weight for each truck or rail shipment. In transportation planning, other considerations are volume (cube), loading (pallet sizes), and consolidation possibilities.

The Transportation Planning Report

The initial planning tool is the transportation planning report which summarizes DRP planned shipments by required ship date and distribution center. The required shipping date will be stated in the DRP time bucket periods but may then require adjustment to the shipping time schedule of the DC. After this adjustment, the summarized planned shipments should be analyzed so as to optimize the transportation plan. If planned shipments for week 2 are 10,000 pounds and week 3 are 60,000 pounds and the best rate is based on full truck loads up to 40,000 pounds, a desirable modification would be to advance 20,000–30,000 pounds from week 3 to week 2. To put this modification in place, a detailed planning report which lists each planned shipment in weeks 2 and 3 will be required. A review of the details will determine candidates which based on weight, cube, and loading requirements can be advanced.

In the above example, the planned shipments under consideration for transfer from week 3 to week 2 must then be reviewed on the MPS of the supplying location to verify availability in week 2 rather than week 3. The master scheduler (or distribution planner) will often plan to have large-volume active items available for shipping "fillers" of this nature. This approach is more practical than adjusting the MPS. Once the detailed modifications are determined, the adjusted planned shipments are stabilized as firmed planned shipments.

The transportation planning reports, both summary and detailed, are tools that assist the distribution function (the distribution planner) in working closely with the manufacturing function (the master scheduler). The report is also useful for projecting freight cost and traffic planning.

The Shipping Schedule

A shipping schedule is required for the shipping department's operation just as a departmental schedule is required in manufacturing. The shipping operation has capacity limits as does a manufacturing operation. Limitations to be considered are stocking space, manpower, truck or rail availability, and rail or shipping doors. An additional consideration will be the receiving and storing capacity of the DC. The shipping schedule should be planned with and made compatible with the DRP output. If there are policies such as full car or truck shipments, they should be implemented by consideration of requirements in advance so as to adjust the DRP plan to the shipping schedule. To make the total system doable, the DRP must still be compatible with the MPS.

When the frequency of shipping to a DC is less than the DRP time buckets, such as weekly DRP requirements and a shipping schedule every 3 weeks, the DRP requirements should be adjusted to the earlier shipping date. As desirable as it is to minimize both inventory and freight costs, it should not be at the expense of customer service. If the DRP planned shipping date is in week 5, but the shipping schedule calls for weeks 4 and 7, the planned shipping order should be advanced to week 4 and firmed.

DRP BENEFITS

Distribution resource planning (DRP) assists in the management of numerous functions within an organization. Marketing through the ability to plan what is needed will show improved customer service levels. DRP is also the tool for Marketing to show problems in the future that must be addressed, as well as assisting in planning promotions. Physical distribution can utilize the DRP system to reduce freight costs, lower inventory levels, and reduce back orders thereby reducing distribution costs, and to improve planning and budgeting. DRP output will assist Finance in cash flow projection, inventory projection, and financial planning. Manufacturing will gain from DRP input to the master production schedule. This input shows more accurate information relating to time-phased requirements.

The above DRP benefits to individual areas within the organization, although positive, are overshadowed by the benefits to the total organization. Just as the MPS is the link that integrates manufacturing and distribution systems, the DRP system is a communication tool that brings distribution and manufacturing operations together through the utilization of similar concepts. All functions are better able to work together through a common understanding of the program. It promotes true teamwork and cooperation throughout the organization.

CASE STUDY

Problem

The original DRP/MPS plan for DC A and plant B is as follows:

DC A
Lot Size = 20, LT = 2 Weeks, Safety Stock = 0

	Week						
	1	2	3	4	5	6	7
Forecast	10	10	10	10	10	10	10
In transit	20						
Projected available balance	10	0	10	0	10	0	10
Planned shipments	20		20		20		

Plant B
Lot Size = LFL, LT = 1, Week Safety Stock = 10

	1	2	3	4	5	6	7
Distribution demands	20		20		20		
Scheduled receipts	20						
Projected available balance	10	10	10	10	10		
Planned order release		20		20			

Two situations arise that affect the plan.

1. The railroad announces that due to siding repairs, there can be no shipments received at DC A in week 5.
2. Due to major electrical problems, the plant must shut down in week 4.

What adjustments must be made to the DRP/MPS plan?

Solution

1. The planned receipt at DC A in week 5 must be moved forward to week 4. The planned shipment in week 3 must be moved to week 2 and "firmed." The distribution demand in week 3 at Plant B is shifted to week 2.
2. The planned order release in week 4 at Plant B must be moved forward to week 3. The planned order in week 2 will be moved forward to week 1. The planned shipments to DC A in week 5 does not have to be rescheduled.

The revised DRP/MPS plan is as follows:

The Revised DRP/MPS Plan is
DC A

	Week						
	1	2	3	4	5	6	7
Forecast	10	10	10	10	10	10	10
In transit	20						
Projected available balance	10	0	10	20	10	0	10
Planned shipments	20				20		
Firm planned shipments		20					
Plant B Lot Size = Lot for Lot (LFL), LT = 1 Week, Safety Stock = 10							
Distribution demands	20	20			20		
Schedule receipts	20						
Projected available balance	10	10	10	30	10		
Planned Order release	20		20				

QUIZ

1. If all distribution centers have level demand, the demand on the central supply center will also be level.
   a. True
   b. False

2. A forecast of a DC is considered a:
   a. scheduled receipt
   b. planned order
   c. shipment release
   d. gross requirements

3. If the master schedule cannot meet the requirements of a DC, the remedy may be to
   I. reduce lot size of the order
   II. increase the safety stock
   III. use safety stock at the DC

   a. I
   b. I and II
   c. I and III
   d. I, II, and III

4. The distribution lead time includes
   I. order release and pick
   II. loading time
   III. unloading time
   IV. in-transit time

   a. I and II
   b. I, II, and IV
   c. I, III, and IV
   d. All of the above

5. To trace specific MPS demand back to distribution centers, you would use
   a. firm planned orders
   b. capacity planning
   c. pegging
   d. projected on hand

6. If the shipping schedule conflicts with DRP requirements, you should
   a. ignore the DRP
   b. change the shipping schedule
   c. move the planned order in closer to meet the shipping schedule
   d. move the planned order farther out to meet the shipping schedule

7. The distribution demands of the central supply master schedule is the sum of the ____________ of the DC centers.
   a. gross requirements
   b. planned order receipts at the DC
   c. planned order shipments to the DC
   d. safety stocks at the DC

8. In a distribution network, safety stock located in a central supply location will cause ____________ inventory than if stored at the DC warehouses.
   a. less
   b. more
   c. the same

9. A planned inventory build up can be stored at
   I. the central supply location
   II. the DC warehouses
   III. an outside warehouse

   a. I
   b. II
   c. I and II
   d. All of the above

10. A distribution requirements planning system differs from a material requirements planning system in that a DRP system
    a. performs time-phased, level-by-level netting
    b. is primarily concerned with finished goods inventories
    c. can consider existing scheduled receipts
    d. can calculate and peg dependent demand

BIBLIOGRAPHY

APICS Dictionary, 7th ed., Falls Church, VA: American Production and Inventory Control Society, 1992.

Martin, A. J., *DRP: Distribution Resource Planning—Distribution Managements's Most Powerful Tool.* Englewood Cliffs, NJ. Prentice-Hall, 1983.

Mosesian, H. J., *DRP: A planning tool . . . and much more.* in American Production and Inventory Control Society 31st Annual International Conference Proceeding, 1988.

Vollman, J. E., Berry, N. L., and Wybark, D. C., *Manufacturing and Control Systems,* 3rd ed. Homewood, IL: Richard D. Irwin, 1992.

9
Job-Shop Execution

Once the management requirements planning (MRP) system has projected a plan, the remaining task is to execute the plan. Through proper consideration of capacity, the evolving plan may be doable, but "doing it"—the execution—can be most challenging, especially in a job-shop environment. MRP was developed as a tool to assist in the complicated dependent time-related material relationships in manufacturing operations. The complicated elements are manifested to the greatest degree in job-shop operations.

A job shop is defined as a group of manufacturing operations where the productive resources are organized according to function and the work passes through in varying lots and routings. The manufactured end item is normally assembled from two or more components which have been fabricated and/or purchased. As reflected in the product structure, each assembly or subassembly is put together with fabricated or purchased components and each fabricated component is made from a purchased raw material. Job-shop execution may also be applicable when the end items are single units rather than assembled. The execution of the MRP plan requires the successful operation of the assembly, purchasing, and fabrication functions. Figure 9-1 illustrates the planning and execution relationships in a job-shop environment. The master schedule drives the MRP which, in turn, plans the purchase orders to suppliers of both raw material and purchased parts. Work orders are planned for fabrication work centers A, B, or C and for subassembly work centers D and E. The final assembly schedule is based on the demand requirements reflected in the master schedule and/or customer orders. This is known as an order-push system as compared to the Just-In-Time (JIT) demand-pull system which is explained in Chapter 11.

ASSEMBLY SCHEDULING

Assembly Environments

The assembly operation in the job shop may be part of make-to-stock, assembly-to-order, or make-to-order environments. In the make-to-stock environ-

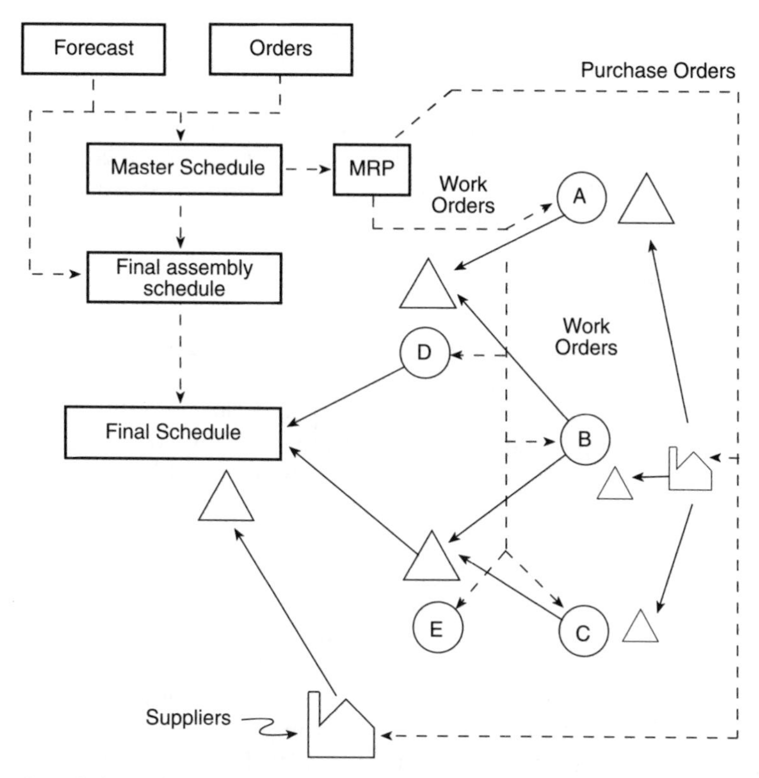

Figure 9-1. Order-push system. [Reprinted with the permission of APICS, Inc., *The Implementation of Zero Inventory/Just-In-Time,*" R. W. Hall, 1986.

ment, there is a level of uncertainty of actual customer demand and, therefore, minimum assembly lead time is desired so as to allow the flexibility to respond to demand change. The assembly lead time is within the frozen time fence of the master production schedule (MPS). Availability of all make-to-stock assembly components is based on the demands of the MPS. The quoted lead time to the customer of make-to-stock "off-the-shelf" items will normally cover the time to pick, pack, and ship the product.

The assemble-to-order environment is based on the availability of all components which will be assembled to specific customer orders. The customer lead time will be based on the assembly and shipping lead times as well as possible adjustments for subassembly and component availability. The availability of components and subassemblies is based on planning or modular bills' demand in the MPS and the customer order backlog.

The make-to-order assembly will have a long customer lead time that may encompass not only the manufacturing critical path (raw material purchase to final assembly) but also include engineering design time. In some make-to-order situations, the lead time is reduced through the stocking of long-lead-time common raw materials and components. The bill of material will be unique for each make-to-order item and will utilize the MPS for material control. The make-to-order environment can have a greater degree of resource fluctuation and, therefore, order promising must consider existing backlogs as well as standard lead times.

Final Assembly Schedule

The final assembly schedule (FAS) is the listing of end items to be assembled. FAS is the tool for releasing assembly orders and the issuing of component pick lists. In some situations, the assembly process may include subassembly operations, finishing, painting, and/or even fabrication not included in the MRP system. These additional activities would be listed in the routing file and must be covered in the lead-time allowance. The FAS must be coordinated with the MPS.

In the make-to-stock environment, the FAS will consist of those items included in the MPS and required for off-the-shelf availability. The MPS and FAS items are identical with the exception of packaging variations stated in the FAS but grouped by end item in the MPS. In assemble-to-order or make-to-order environments, the FAS will be stated by customer order. The end item will be configured to the customer requirement and will consist of modules and components that are master scheduled at the next bill of material level. Although the FAS items are distinct from the MPS items, they must be consistent.

The make-to-stock orders will be planned and released based on desired inventory levels, lot sizes, and lead times. The scheduler must verify both parts availability and assembly capacity prior to order release. Once the verification has been accomplished, additional adjustments may be required for load leveling and coordination with the MPS. Table 9-1 illustrates an example where the demands of the distribution system with requirements based on customer demands can cause instability with the MPS when driven by the time phased order point of distribution demands. In week 1, an assembled product is planned for release in weeks 2, 4 and 6 and is consistent with the MPS plan. Actual shipments in week 1 is 15 rather than 10. This causes the planned assembly releases in weeks 4 and 6 to be advanced to weeks 3 and 5. In week 2, the actual shipments are 2 rather than 10. This causes the planned assembly releases in weeks 3 and 5 to be shifted back to weeks 4 and 6.

If there are not too many items to control, the master scheduler can decide to use the safety stock and stabilize the MPS by firming the planned assembly orders. With many items to control, the stabilization can be affected by

Table 9-1. FAS/MPS relationship.

Central Warehouse Requirements (TPOP)
Lot Size = 20, Safety Stock = 10, Lead Time = 1 Week

		Week						
		1	2	3	4	5	6	7
Forecast		10	10	10	10	10	10	16
Scheduled receipt		20						
Projected available balance	10	20	10	20	10	20	10	20
Planned assembly order release (MPS)			20		20		20	
Week 1 Shipments = 15								
Forecast			10	10	10	10	10	10
Scheduled receipt				20				
Projected available balance		15	5	15	25	15	25	15
Planned assembly order release (MPS)				20		20		
Week 2 Shipments = 2								
Forecast				10	10	10	10	10
Scheduled receipt				20				
Projected available balance			13	23	13	23	13	23
Planned assembly order release (MPS)					20		20	

extending the firm time fence in the MPS. This improves stability but reduces flexibility.

The FAS for assemble-to-order and make-to-order items must be coordinated with the MPS to assure material availability for assembly. The customer order will be configured by specific modules and options which have been master scheduled. This temporary assembly bill is then utilized as the assembly order in the FAS. The FAS requirements then reduce (consume) the affected modules and options. If the master schedule is a two-level MPS, the second level (Level I) items' forecasts will be reduced by the FAS requirements. As with make-to-stock assemblies, parts availability and capacity must be considered. Scheduled completion dates (i.e., promised delivery) will be dependent on backlog and, in the make-to-order environment, the reliability of open work order and purchase order on time delivery.

PURCHASING

Present-day philosophy tells us to treat suppliers as an extension of the manufacturing operation. This approach calls for the scheduling of purchased parts and raw material delivery for the job shop just as we are to schedule fabrication and assembly operations. A customer's understanding of the suppliers' operations is not normally as detailed as the data in the manufacturing routing and work center files. However, some of the MRP manufacturing concepts can be carried over to purchasing operations. For example, a supplier can be assisted in resource management for a given customer if the customer will supply the MRP planned orders for parts that are single sourced to the supplier. This MRP output is to be for information purposes only and does not constitute an order nor commitment. The controlling elements considered in internal scheduling should also be considered in the purchasing operation.

Lot Sizing

Basing the lot size of a purchased part on the economic order quantity (EOQ) formula with the ordering cost in place of setup does not consider all cost aspects. If the part is a commodity that the supplier maintains in stock, the cost of packaging and transportation is affected by the lot size. If the part is produced specifically for the customer, the supplier's manufacturing setup costs along with the packaging and transportation are the proper considerations. Most suppliers will cover their lot size costs by promoting larger lot sizes through quantity discounts. *Production and Inventory Management* (2nd ed) by Fogerty, Blackstone, and Hoffman gives an excellent example of tabulating total costs of purchasing based on various lot sizing techniques. Lot for lot, EOQ, package size, and price break quantity are compared (price break quantity is the least expensive in the example).

Lead Time

The MRP planning lead time for a purchased item must include the following:

1. Order preparation time
2. Supplier lead time
3. Transit time
4. Receipt to stock time

The due date for MRP planning is the date to stock, whereas the due date to the supplier will be the date for delivery to the plant. Reducing the planning lead time will not decrease work-in-process, but will reduce the critical path, making the system more flexible and able to operate to a shorter forecast horizon.

Lead time can be reduced through agreements with suppliers that commit the customer to "buy" capacity or to take responsibility for long-term commitments to their suppliers. An example of buying capacity would be an arrangement with a foundry for the purchase of 2000 castings a week. The foundry's backlog (and, therefore, stated lead time) might be 14 weeks, but the actual manufacturing time could be as little as 4 weeks. A blanket order (with a dummy part number, if necessary) could be placed for 28,000 castings, but specific castings would be ordered based on a 4-week lead time. This system would require releasing control to maintain an even flow of 2000 castings per week. This can be accomplished through the use of input–output control similar to that used for a work center. The supplier (like a work center) would have a desired backlog of work; in this case, 8000 castings.

An example of reducing lead time by taking responsibility for a supplier's commitment would be an assembled product with a supplier's lead time of 16 weeks. Analysis shows that the 16-week lead time consists of 3 weeks assembly time and 13 weeks of unique purchased parts' lead time. Further analysis shows that the unique parts are only 20% of the total cost. An arrangement can be made to publish the MRP output of the assembled product to the supplier with the stipulation that the supplier can order out 20–26 weeks for the unique parts and that the customer will accept responsibility if they are not used. The supplier, in turn, will reduce the lead time for the assembly to 4 weeks.

Single Source

The lead-time reduction techniques are based on the assumption that the part is single sourced (supplied by only one supplier). Single sourcing is based on a relationship of confidence between both parties. In this situation the customer has a choice of the single-source supplier as opposed to a sole-source supplier, where, due to patents, exclusive tooling, or other limitations, the customer has no choice. The confidence factor for single sourcing includes the knowledge that the supplier has the necessary capacity and flexibility to meet the needs of the job-shop manufacturer. Other considerations are the labor relations climate and the financial stability of the supplier. If the supplier is to be considered an extension of the manufacturing operation, single sourcing by part number is desirable.

With a multiple-source arrangement, a source decision is required for each purchase. This is similar to each fabricated part having three alternate routings and one of the routings having to be chosen at the time of work order release. This is not impossible, but it is not practical. Such decisions are required for new parts, purchased or fabricated, but, once in place, the best routing or supplier should be utilized.

Supplier Certification

The formality of establishing a single source may involve the certification of the supplier. Certification requirements are established by each customer and may consist of such approaches as ISO 9000 qualification, Malcolm Baldridge award application, attaining certain quality levels, or passing a detailed evaluation program designed by the customer. There have been many success stories published, but there have also been many failures. It does not help supplier–customer relationships when the customer who is known to be anything but efficient sends his experts in to "help" the supplier.

Certification at the part number level is based on the suppliers consistently meeting the quality standards of the part. A certified part will not require incoming inspection, but will be monitored based on a sampling process. The sampling may take place at the supplier's facility rather than after delivery. This is a positive example of working together rather than being adversaries.

Purchase Order Control

Just as the fabricating work center is controlled with a dispatch list, supplier control can be maintained by a supplier due date list of all open part numbers. It is not practical, even if possible, to monitor the progress of a purchased part in the same detail as in the customer's internal operations. The key to successful communications is for the customer to maintain valid due dates on the list and the supplier to feed back any potential problems in meeting the due dates. Some systems utilize pegging routings to list not only the quantity and due date of each order, but how many units of the order are required for specific orders as compared to planned stocking.

Electronic data interchange (EDI) is the utilization of computers to communicate and exchange documents between customers and suppliers. This paperless system includes purchase orders, releases, delivery notification, and invoices. Besides the reduction in paperwork, there are the advantages of reducing transaction time and quick response to change. Implementation of EDI has been no easy task, due to problems relating to differing formats for computer data storage and differing data-base management programs. As with supplier certification programs, EDI has the potential to significantly improve purchasing operations, but the implementation phase can be most painful to suppliers.

FABRICATING SCHEDULING—INFINITE LOADING

The execution of fabrication operations in the job shop is normally based on recommended MRP orders. In some situations, the fabrication requirements are generated by the actual rather than anticipated needs of the assembly schedule. This is called a "pull" approach as compared to the "push" of the

MRP system and is a concept of Just-In-Time manufacturing. The goals of the scheduling system are as follows:

1. Complete fabrication by the due date with
2. minimum lead time and
3. maximum machine utilization

Lead time and machine utilization goals conflict when, due to random fluctuations of work arriving at work centers, favorable machine utilization is achieved through large queue allowances. The large queue allowance increases the lead time, which, in turn, increases work-in-process. With infinite loading systems, capacity has been addressed with the capacity requirements planning (CRP) review of the MRP, but this is no guarantee that work scheduled for a given period will match the capacity of the machine or work center, especially on a daily or hourly basis. Problems can result with too many jobs scheduled at one time. Another problem can be when the scheduled work matches the capacity, but is not available due to problems with previous operations. The operation-by-operation schedule is based on the MRP requirements, the planned setup and run times reflected in the routing file, and the queue, move, and wait allowances of the system. Backward scheduling calculates the start and due dates for each operation by starting with the order due date and computing backward. Forward scheduling starts with the planned order release date and computes each operation's start and due dates from first to last operation.

The Dispatch List

The dispatch list is a priority schedule for each work center that is normally published daily or in real time and lists the work load with a visibility of 2–5 days. Table 9-2 is an example of a dispatch list for a turret lathe operation.

Table 9-2. Dispatch List for Turret Lathe Department 381—Work Center C

Part No.	Work Order	Quantity	Operation Start	Operation Finish	Setup Time	Run Time	Location
10704	9702		7/8	7/9	1.0	5.0	381
09728	9707		7/8	7/9	.5	4.5	381
03496	9671		7/9	7/11	1.5	7.0	354
21245	9712		7/10	7/12	1.0	6.5	371
36712	9730		7/10	7/12	.5	3.5	360
19792	9726		7/10	7/12	1.0	6.5	360

If work center C is a one-machine work center and is scheduled one shift per day with a capacity of 7.5 standard hours per shift, the output for 5 days would be 37.5 hours compared to the scheduled 38.5 hours. This is a doable schedule if the machine does not break down and the last four jobs get to the work center on time.

The due dates on the dispatch lists are important, as most hard-copy shop orders no longer carry due dates. Experience has shown that for a variety of reasons, the due dates at time of order release cease to be valid during the lifetime of the shop order. The daily published dispatch lists have at times been confusing to the shop floor due to updating problems such as the job that completes 20 minutes after publishing or the priority that changed from yesterday and the job now being set up is no longer needed for the next 6 days. On-line CRT display of the dispatch list has reduced much of the confusion.

Prioritizing Rules

Whether employing finite or infinite loading techniques, rules of prioritizing orders must be established. Common priority rules are as follows:

1. Due date control based on either earliest operation due date or start date. If all work is available as scheduled and there are no capacity problems at the work center, this is the rule of choice.
2. Schedule the jobs in order of least total slack time remaining. Slack time is determined by subtracting the remaining setup and run times from the time remaining to due date completion. Order A and Order B both are due in 8 working days. Order A has 2 days of setup and run time remaining and, therefore, 6 days of slack time. Order B has 5 days of setup and run time and only 3 days of slack time. Order B will be scheduled first.
3. Schedule the jobs with the shortest processing times first. This rule will cause lower work-in-process (WIP), but is only valid when all jobs are running late. It should not be used if it does not consider due dates.
4. Schedule the job first with the lowest critical ratio. The critical ratio is the order time remaining to due date completion divided by the manufactured lead time remaining to completion. Manufactured lead time includes queue and move allowances. The job due in 6 days, but with only 4 scheduled days remaining, has a critical ratio of 1.5 and is running ahead of time. A job with 5 days remaining for both completion and scheduled time has a critical ratio of 1.0 and is considered on time. A job with 7 days of scheduled work, but is due in 5 days, has a critical ratio of 0.7 and is late. The priority of the three jobs should be

1. Critical ratio 0.7
2. Critical ratio 1.0
3. Critical ratio 1.5

If all jobs to be scheduled have critical ratios of less than 1.0, the plant is in trouble.

There are other variations of priority rules, but some such as first in, first out or longest processing time conflict with the first goal of the system, completing the fabrication by the due date. Judgment is required in the execution of the system. There is no point in running a job first, no matter what the priority rule, if it is going to sit in queue for 2 weeks at the next operation.

Shop-Floor Control

Maintaining the status of the WIP is based on the shop-floor control system. In job-shop fabrication operations, shop-floor control is maintaining the detailed operation status and quantities of each open shop order. The data are used for prioritizing dispatch lists, output for CRP control, cost system management, and shop performance measurements. Data integrity is most important and has not always been easily achieved. Recent improvements have been brought about by increased operator training, technological innovations such as bar coding, and the JIT approach of simplification by only collecting what is really important. Real-time data collection in a distributed data processing mode collects information on site which will be utilized for local control and scheduling. Required data are communicated to the central computer controlling the total WIP operation.

Shop-floor data are utilized for measuring labor performance, machine utilization, and quality, all of which are important, but do not measure the control system. Examples of control system performance measurements and their targets are listed in Table 9-3. Other critical measurements such as customer service are based on data collected from sources other than the shop floor.

The success of an infinite scheduling system is best measured by shop order completion on time and is dependent on releasing work within capacity constraints, prioritizing rules, and adherence to schedule.

Table 9-3. Shop-Floor Performance Measurements

Performance Measurement	Target
MPS orders on time	95–100%
Fabrication shop order on time	95%
Shop order lead time	At or below planned lead time
WIP inventory turns	At or above predetermined standards

Shop-Floor Management

The best prioritizing and dispatching system in the world will not succeed if the manufacturing environment causes conflicting goals. The pressure on the foreman to create favorable labor variance numbers and incentive systems that cause operators to want to "cherry pick" favorable jobs run contrary to good scheduling. Another factor to be resolved is who is going to control the schedule. The foreman knows the people, the machines, and the tooling. The planner in production control understands the requirements of the order. The most reasonable compromise is for production control to do the scheduling, but for the foreman to be allowed adjustments within predetermined limits. An advantage of an infinite oriented dispatch list is that the foreman can easily understand the logic. A common situation is when setup times can be substantially reduced (or in some cases eliminated) by combining jobs. The combining of the jobs may cause running an order out of sequence. The desired result can happen when the foreman has enough system knowledge and understanding to know when such variation of sequencing will not have an adverse effect on the on-time completion of the part. In other words, will there be adequate slack time in the balance of the run schedule and will the system accommodate it?

Routing the job-shop fabrication operations are planned by the manufacturing engineers and have a strong impact on the success of a scheduling system. The more complicated the product routing, the greater the chance of random fluctuations and bottleneck problems. The key to WIP management is individual queue control monitored by input–output control techniques. The number of queues to be controlled are a function of the number of operations to be scheduled and monitored. If operations are combined and the number of managed queues reduced, the task of WIP management becomes easier and the change of successful product flow increases.

In some situations, the setup times for a number of operations may be excessive and therefore require large lot sizes. In this situation, the shop order may not have to be completed for each operation before scheduling the next operation. This method of scheduling is called overlapping. Most software systems allow for overlap allowances in the routing file and will schedule based on the allowances. Table 9-4 illustrates the effect of an overlapped schedule.

FABRICATION SCHEDULING—FINITE LOADING

Scheduling fabrication operations in a job shop with finite loading techniques differs from infinite loading in that capacity is addressed and the system will

Table 9-4. Overlapping Schedule

Conventional Job-Shop Schedule

Operation	Queue, Setup, and Run Time	Scheduled Start	Scheduled Completion
10	40 hours	Day 1	Day 5
20	48 hours	Day 6	Day 11
30	32 hours	Day 12	Day 15

Overlapping Job-Shop Schedule

Operation	Queue, Setup, and Run Time	Overlap Allowance	Scheduled Start	Scheduled Completion
10	40 hours	80%	Day 1	Day 5
20	48 hours	50%	Day 2	Day 7
30	32 hours		Day 5	Day 8

not allow the work load to exceed capacity. Real-time forward finite loading may not only consider the capacity of the work center but also the availability of the shop order, manpower, and tooling. The system will collect and monitor the data and then load based on predetermined priority rules.

Finite Loading Problems

Problems (or as the Japanese would say, opportunities) with finite loading is that the results may conflict with MRP requirements. In other words, infinite loading is compatible with MRP requirements, but may not be realistic while finite loading will be realistic, but may not be compatible with MRP requirements. A second problem that can arise is that a finite loaded order based on the priority rules may make sense based on the scheduled work center, but not if work load down the line is considered. If prioritizing is by operation due date, item A (due for completion May 5) will be loaded ahead of item B (due for completion May 6). If item A's next operation is at a work center with a 6-day queue and item B's next operation is at an open work center, item B should run first. Another problem facing tie-breaking situations in finite scheduling is what is the end use of the scheduled items. Item C has an MRP required due date of November 9 and item D has an MRP required due date of November 10. Prioritizing rules would give precedence to item C over item D, which is reasonable unless item C is being fabricated for stock while item D is needed for an assembly order.

Finite scheduling is accomplished with stand-alone systems calling for sophisticated data control and calculations. It is most important that these systems be integrated with the planning system (i.e., MRP requirements). Finite loading must be continually compared to planning requirements in order to make expedient adjustments to either planning or execution. Because of the rather complicated planning logic with finite loading, foremen are often not comfortable with the dispatch lists. The logic is not transparent. Two scheduling techniques that assist in addressing finite loading problems are operation sequencing and bottleneck scheduling.

Operation Sequencing

Operation sequencing is a simulation approach in which the initial calculation generates a short-term plan at each work center based on finite scheduling priorities. These plans are then projected to show completion times and simulates queues. The simulation continues by moving the job to the next work center and then the second day's run is simulated. Simulation can predict both potential overloaded and idle work centers and therefore assist the planner (dispatcher) in making priority adjustments to short-term plans prior to the extension of the simulation-based schedule into real-time dispatching. Operation sequencing should be generated at least daily or on a shift-by-shift basis. Table 9-5 is an example of operation sequencing of three parts which based on due date priority scheduling at the gateway operation will all complete late due to queues at two down-line work centers. Table 9-6 shows the results of adjusting priorities.

Table 9-5. Operation Sequences

Item A				Item B				Item C			
Operation	Work Center	Run Hours	Due Date	Operation	Work Center	Run Hours	Due Date	Operation	Work Center	Run Hours	Due Date
10	350	8	12/1	10	350	8	12/2	10	350	8	12/3
20	370	8	12/3	20	360	8	12/4	20	360	8	12/5
30	390	8	12/5	30	370	8	12/6	30	370	8	12/7

Based on one 8-hour shift–7 days/week, each Work Center's Gantt Chart (Where "X" is backlog of other prioritized items) is

	12/1	12/2	12/3	12/4	12/5	12/6	12/7	12/8
WC 350	A	B	C					
360			X	X	X	B	C	
370		X	X	X	A		B	C
390						A		

Note: Due to the backlogs at Work Centers 360 and 370, all three jobs complete one day late.

Table 9-6. Operation Sequencing After Adjustments

Reversing the gateway schedule of items A and B shows that while item A will still be late, items B and C will complete on time.

	12/1	12/2	12/3	12/4	12/5	12/6	12/7
WC 350	B	A	C				
360		B	X	X	X	C	
370		X	X	X	A	B	C
390					A		

Bottleneck Scheduling

Focusing and controlling resources when capacity is equal or less than demand is called bottleneck scheduling or constraints management. Identification of the bottleneck machines or work centers can be difficult due to large queues that may form at nonbottleneck centers because of bottlenecks at previous operations. JIT philosophy says to identify and eliminate the bottleneck. If elimination of bottlenecks is possible, work flow can be achieved through implementation of JIT pull scheduling techniques. In most situations, the bottleneck(s) cannot be eliminated and must be dealt with through a planning system that will optimize the utilization of the bottleneck.

Optimized product technology (OPT) is a philosophy that addresses the goal of never shutting down the bottleneck and having it run the highest priority items. (OPT originally was the acronym for optimized production timetables.) The bottleneck is finite loaded with the queue control based on back scheduling to the bottleneck with MRP logic. In planning material arrival at the bottleneck, overlap scheduling may be called for in order to optimize the bottleneck schedule. The most easily controlled bottleneck is when that bottleneck is the gateway (first) operation.

As the material is processed through the bottleneck, the lot size may be reduced (i.e., split) and forward scheduled through the nonbottleneck operations. OPT philosophy is that nonbottleneck work center scheduling should not be concerned with setups and, therefore, a lot-for-lot technique of lot sizing will increase the product flow rate and decrease work in process.

Software systems that address bottleneck scheduling are still in the developmental stages.

CASE STUDY

Problems

1. A custom furniture manufacturer operates in a make-to-order environment. The normal lead time is 8 months with manufacturing time equal

to 3 months and fabric procurement lead time of 5 months. Wood is the other major supply item and is available "off the shelf" from a local lumberyard. There are 200 fabric items in the sample book with 40 items representing 80% of the usage. Competition is now quoting 4 months. What action should the manufacturer take to be competitive relative to lead time?

2. On Day 4, three items are to be loaded to a work center. The detailed data are as follows:

Item	Operation Due Date	Setup and Run Time Remaining	Manufacturing Lead Time Remaining
A	Day 8	1	2
B	Day 10	4	7
C	Day 11	6	7

What would be the loading sequences based on operation due date, least total slack, and critical ratio priority rules?

3. The infinite loading of an intermediate work center with 16 hours of capacity calls for the following 27 hours of production (based on same due date prioritizing):

Item	Setup and Run Time
A	8 hours
B	9 hours
C	10 hours

There is no alternative other than to move one of the three items off the schedule. What might the dispatcher consider in order to make the best decision?

Solutions

1. The manufacturer can reduce lead time from 8 to 3 months by moving from a make-to-order environment to an environment similar to assemble-to-order. This will require only the stocking of fabrics as wood is available at the lumber yard. The 40 popular fabrics would be stocked and controlled at the MPS level based on forecasted usage. The remaining 160 fabrics (20% of the business) would be offered only as "specials" with an 8-month lead time.

2. The criteria required for the priority rules are as follows:

AEM	Operation Due Date	Time to Completion	Setup and Run Time	Slack Time	Manufacturing Lead Time Remaining	Critical Ratio
A	Day 8	4	1	3	2	2.00
B	Day 10	6	4	2	7	0.86
C	Day 11	7	6	1	7	1.00

The loading sequence is based on

Operation due date A–B–C
Least total slack C–B–A
Critical ratio B–C–A

3. Whereas the due date prioritizing rule gives all three orders equal status, other prioritizing rules might assist in the decision of which item should be taken off the schedule. Factors for consideration are as follows:
   a. The amount of total slack time left for each order.
   b. The ratio of time to completion compared to planned manufacturing time (critical ratio).

Other considerations might be the end use of the items which would be determined by pegging or the review of the work center queues of remaining operations for each item (simulation).

QUIZ

1. The assumption that a shop order will be completed by the MRP generated due date always requires
   I. capacity planning
   II. finite loading
   III. infinite loading
   IV. operation sequencing

   a. I
   b. I and II
   c. I and III
   d. IV

2. Multiple jobs may be scheduled at the same time when __________ is used.
   a. finite loading
   b. infinite scheduling

3. Finite scheduling assumes infinite capacity.
   a. True
   b. False

4. Bottleneck scheduling assumes
   I. one or more bottlenecks exists
   II. all operations in use be finite loaded

   a. I
   b. II
   c. I and II
   d. Neither I nor II

5. Operation sequencing
   I. is a short-term planning technique
   II. looks ahead to the next operation

   a. I
   b. II
   c. I and II
   d. Neither I nor II

6. Finite scheduling
   I. considers capacity at the work center
   II. loads only to capacity levels
   III. prioritizes the work load

   a. I
   b. II
   c. III
   d. All of the above

7. With infinite loading, you
   I. assume available capacity
   II. risk multiple jobs scheduled at the same time
   III. load only to capacity levels

   a. I
   b. I and II
   c. I and III
   d. All of the above

8. The basic control document for job shop work-in-process is the
   a. labor ticket
   b. work order
   c. material requisition
   d. drawing

9. Operation scheduling for the dispatch list can be calculated by either forward or backward scheduling.
   a. True
   b. False

10. If the rate of unforeseen events are excessive, you must use
    a. operation sequencing
    b. finite loading
    c. infinite scheduling
    d. safety stock

BIBLIOGRAPHY

Berger, G., *Solving job shop planning and scheduling with MRP II,* in American Production and Inventory Control Society 31st Annual International Conference Proceedings, 1988.

Donovan, R. M., *Enhancing the schedule execution capability of MRP systems.* in American Production and Inventory Control Society 35th Annual International Conference Proceedings, 1992.

Fogarty, D. W., Blackstone, J. H., Jr., and Hoffman, T. R., *Production and Inventory Management.* Cincinnati: South-Western Publishing, 1991.

Hall, R. W.; *The Implementation of Zero Inventory/Just-In-Time,* 1988.

Thompson, M. B., *Computer simulation drives innovation in scheduling,* APICS, *The Performance Advantage,* February, 1993.

Vollman, J. E., Berry, N. L., and Wybark, D. C., *Manufacturing and Control Systems,* 3rd ed. Homewood, IL: Richard D. Irwin, 1992.

10 Process Manufacturing Execution

Process manufacturing is best understood by defining the nature of the material routing and the related manufacturing activities. Whereas job-shop manufacturing involves differing material flow routes through functional departments, process manufacturing has the characteristic of fixed routings. Both process and repetitive manufacturing have fixed routing characteristics, but they differ. Process manufacturing will add value to few materials and produce end items through mixing, forming, or separating operations, whereas repetitive manufacturing consists of fabrication and assembly operations. Table 10-1 lists examples of process, job shop, and repetitive manufacturing products. The goal of repetitive manufacturing is to achieve a process flow similar to process manufacturing and it is one of the essential ingredients of Just-In-Time (JIT), which is covered in Chapter 11.

Table 10-1. Manufacturing Variations

	Products
Process manufacturing	Plastic strap
	Paint
Job-shop manufacturing	Made-to-order machines tools
	Low-volume generators
Repetitive manufacturing	Automobiles
	Television sets

Process manufacturing is characterized by shallow bills of material as well as fixed routings. There are differing characteristics within process manufacturing in that some products will be made in a continuous flow, which, in turn, will allow small work-in-process (WIP) inventories and short lead times, whereas other products will be produced in batches and may be subject to lot size considerations. Dependence on setup considerations for specific operations in the process routing may call for maintaining inventories at differing stages of the process and will result in larger WIP inventories while still being produced in a relatively short lead time. Continuous process scheduling will be capacity-driven with the only material consideration being raw material and without any shop-floor or priority control required. Scheduling considerations for batch process manufacturing does call for additional planning and control. Maintenance planning is especially important with continuous processing, as a breakdown will immediately shut down the entire line. The planning function may or may not be material requirement planning (MRP) or master production schedule (MPS) generated. Figure 10-1 is a repeat of the conventional MRP II closed-loop process illustrated in Chapter

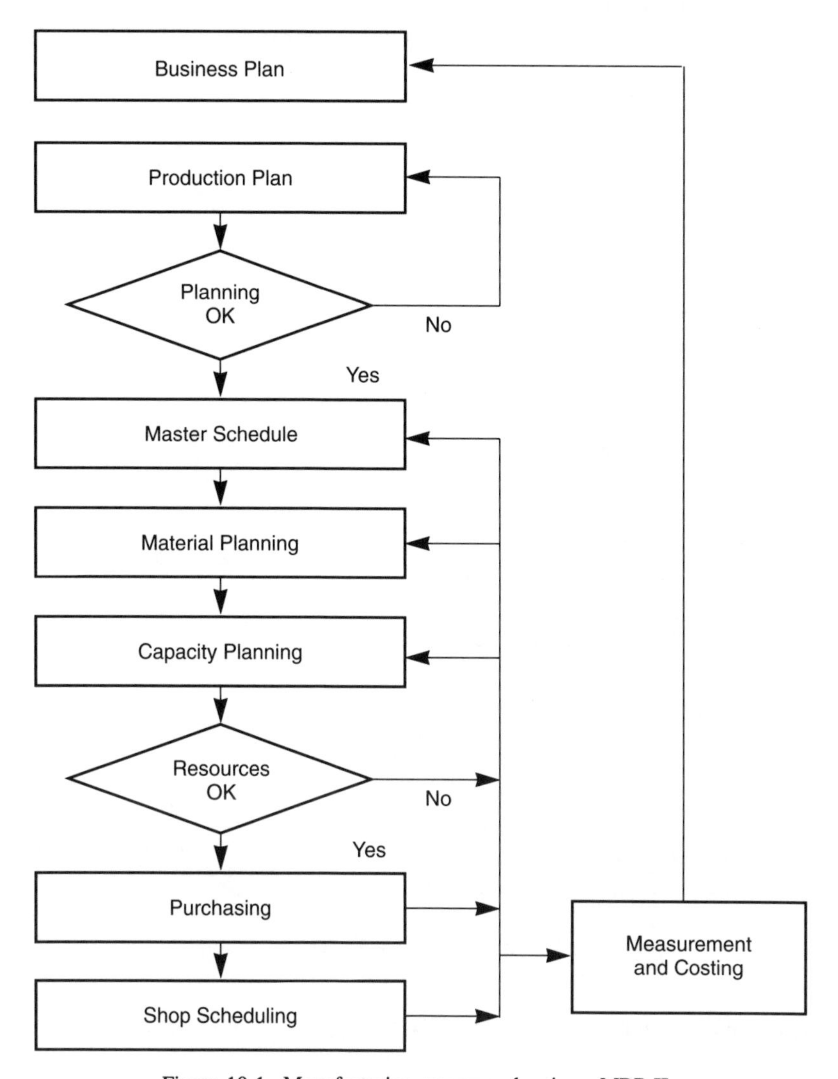

Figure 10-1. Manufacturing resource planning—MRP II.

1 (Fig. 1-3). A manufacturing operation that controls raw material by an MRP system will use this process with emphasis placed on capacity planning.

When material control is maintained with a separate system and the MPS is utilized for shop scheduling, the closed-loop process is depicted in Figure 10-2.

In some environments, the production plan may be the driver of shop scheduling with process capacity the major consideration. The process structure is emphasized in planning and control. This scheduling system is known as process flow scheduling (PFS).

This chapter details the various scheduling approaches utilized in process manufacturing.

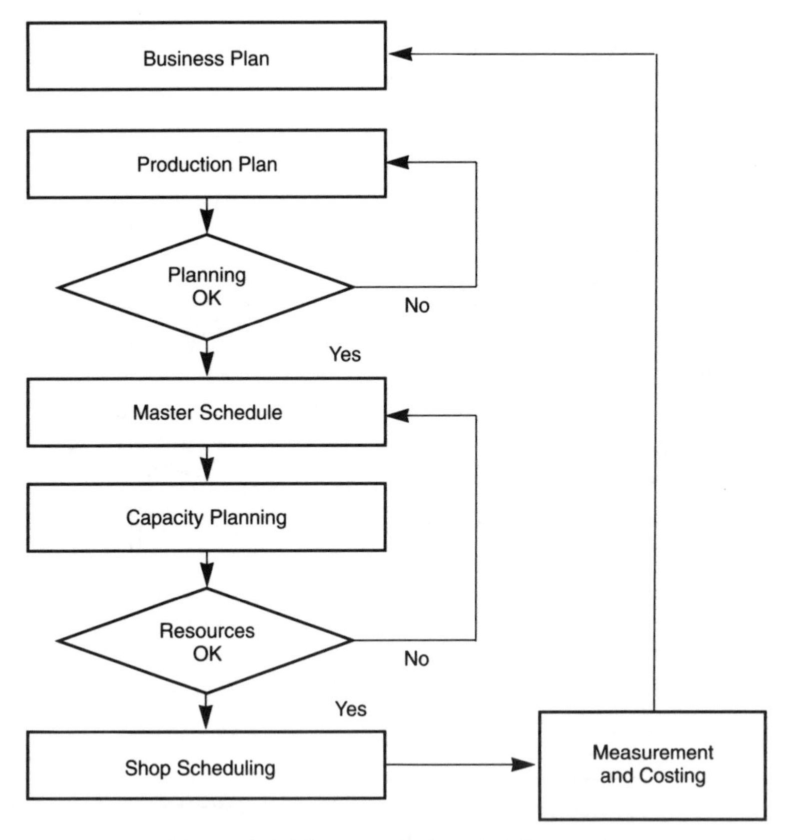

Figure 10-2. Master production schedule control.

CAPACITY-DRIVEN MASTER PRODUCTION SCHEDULING

The rule of "never lie to the master production schedule" is especially true in process manufacturing capacity considerations. Once the schedule becomes operational, the production flow is not easily adjusted by priority changes, alternate routings, or expediting. With fixed routings, there is little flexibility, especially with continuous-flow operations, and, therefore, detailed and formal capacity planning is necessary. The level of detailed planning is further increased with the number of operations to be scheduled. If the process is multi-operational and one operation is the bottleneck, that operation's capacity will control the MPS.

The MPS planning horizon may be dependent on raw material procurement lead times or it may be greater if planning for seasonal finished goods. Often the demand during the high season of a product is greater than production capacity and, therefore, finished goods inventory buildups must be planned in the off season. With short manufacturing lead times, the firm part of the MPS is similar to the final assembly schedule in job-shop manufacturing in that the schedule is fixed except that in process manufacturing, part shortages are not a problem. Machine breakdowns and quality issues are process manufacturing's major problems. In make-to-order environments, MPS demand is order-entry-driven and tends to be fixed based on delivery promises. In a make-to-stock situation, the MPS initial demand will be forecast-driven, but may call for continuous adjustments due to actual demand placed on the distribution system.

The degree of flexibility allowed in the MPS is dependent on raw material availability and, more often, on capacity. Negative influences for flexibility are long setups and bottleneck operations. Positive influences for flexibility are that often raw materials are common to most finished goods and that the system can store WIP buffer stocks at various work centers in batch process manufacturing.

In continuous processing when MRP is not required (except for procurement of raw materials), the MPS will generate production orders, set priorities, and control the dispatch list. With relatively few products to be scheduled, these systems may be manual rather than computerized.

When routings are fixed, production flow rates are similar and a few raw materials used for many finished goods (inverted bills of material), the MPS may be controlled at the raw material level. The fewer the number of items to forecast, the greater the accuracy. An example of this approach would be a laminations manufacturer with a press, anneal, and pack process utilizing 19 purchased steel sizes to produce 2200 end items.

THE ROLE OF MRP

Continuous process manufacturing with fixed routings, shallow bills of material, and short manufacturing lead times do not require the time-phased dependent demand logic of MRP for scheduling and WIP control. If there are finished goods calling for a high degree of raw material commonality or raw material with long procurement lead times, MRP systems can be useful in purchasing management. When process manufacturing is in a batch mode, with four or more bill of material levels, longer manufacturing lead times, routing variations, or work center dependent lot sizes, planning and scheduling complications can be similar to those of job-shop environments and MRP logic is needed.

When an MRP system is used only for raw material control, it may be driven by a formal MPS program or a production plan. The production plan is less detailed than the MPS, but will supply adequate data for purchasing control. The more detailed information normally found in the MPS will be supplied through specific production orders when scheduled within the manufacturing lead times. The production orders will be based on customer orders or stocking requirements. The MRP system of raw material control will call for bill of material and inventory file data and be based on conventional time-phased gross-to-net logic.

When the manufacturing process consists of multiple operations with the operations having different setup times and running speeds, batch control and WIP inventory levels are needed. For example, operation 1 might have a setup time of 4 hours and a running rate of 10,000 units per hour, whereas operation 2 has a setup time of 10 minutes and a running rate of 500 units per hours. This approach requires additional bill of material levels, increased WIP inventory, and a more complicated planning system (MRP). Although this process runs contrary to the concept of continuous flow, the nature of the manufacturing system makes it necessary for meeting customer requirements and machine utilization. Table 10-2 lists three finished goods (steel strapping) and their process routing. All three are produced from the same 40-in. hot-rolled carbon steel roll.

Table 10-2. Process Routings

	Product		
Routing	A—½-in. width 0.020 gauge	B–5/8-in. width 0.020 gauge	C—1¼-in. width 0.020 gauge
1	Cold roll carbon steel "D" to 0.020	Cold roll carbon steel "D" to 0.020 in.	Cold roll carbon steel "D" to 0.020 in.
2	Split roll to 5 × 8 in.	Split roll to 4 × 10 in.	Split roll to 2 × 20 in.
3	Slit 8-in. roll to 16 ½-in. bands	Slit 10-in. roll to 16 5/8-in. bands	Slit 20-in. roll to 16, 1¼-in. bands

If there are no lot size nor line balancing constraints on the rolling, splitting, and slitting operations, the strap bands can be produced in a continuous flow and be defined by a two-level bill of material.

A more realistic and complicating situation would list the following constraints.

1. The rolling mill calls for a lengthy setup time and, therefore, a minimum run size of 120,000 pounds. This quantity represents 3 weeks of 0.020 gauge requirements.
2. The roll splitter does not require a minimum run size, but only operates at a fraction of the running speed of the strap line slitter.

These constraints require stocking levels after both the rolling and splitting operations. The bill of materials must be expanded to four levels as shown in Figure 10-3. Dependent demand material control for stocking parts D, E, F, G, and H would be managed by the MRP system.

In continuous process manufacturing, the scheduling priorities are established at the production planning or MPS level with little if any shop-floor control. The production order is released to the predetermined process line based on previous capacity and material availability checks. The scheduling considerations for both continuous or semicontinuous (batch) processes are as follows:

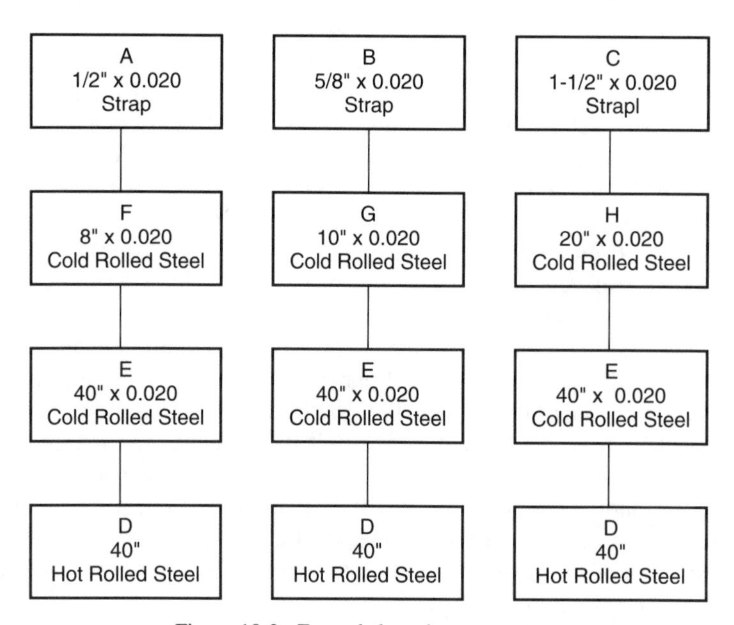

Figure 10-3. Expanded product structures.

1. Prioritizing to meet customer order or MPS requirements
2. Line balancing
3. Operation efficiency
4. Minimizing major setup through family groupings or color sequencing
5. Time-critical products such as when aging, fermentation, or baking reaction times are tightly controlled

Shop-floor data collection in continuous processing environments will be for quality and cost control purposes rather than WIP planning and control. Lead-time measurements will be in minutes or hours compared to days in batch operation. Bar coding and electronic control techniques can be most useful.

In semicontinuous process scheduling, the priority considerations and goals are similar to continuous processing but are more complicated and difficult to achieve. More detailed planning and shop-floor control systems are required. There may be combination production lines with common processes rather than dedicated flow lines. One work center may feed a number of work centers each with differing operations. Each work center or operation may require a specific schedule or dispatch list based on MRP planning and prioritized with finite scheduling procedures similar to those used in job-shop environments. This can include techniques such as simulation on a shift-by-shift basis (operation sequencing).

If there is a bottleneck operation (and there usually is), the bottleneck is the controlling factor in capacity management. The MPS or production flow must be realistic relative to the bottleneck operation. The bottleneck schedule is paramount with all before operations backward scheduled to meet the demands of the bottleneck schedule. Operations after the bottleneck can be forward scheduled with no priority control required if there is excess capacity. Shop-floor WIP should be continuously monitored, as bottlenecks can change with product mix and the work center that was considered to have excess capacity may suddenly be overloaded.

Although the manufacturing process may be semicontinuous and call for sophisticated systems such as MRP planning, capacity planning, and finite scheduling with operation sequencing, the implementation and maintenance of the systems may not be that difficult. If the products and manufacturing operations are not too numerous and the lead times short, it may be practical to operate a manual system. MPS and MRP planning may be possible through the use of a spreadsheet. The JIT concepts of product and process simplification, level work-load flow, and pull system execution where possible should be part of the scheduling strategy.

PROCESS FLOW SCHEDULING

Process flow scheduling (PFS) is presented as an alternative to MRP. With PFS, the scheduling calculation is guided by the process structure rather than the bill of material. Greater emphasis is placed on process run rates and resources rather than the product structure and routing. The planning process is driven by a production plan rather than MPS and can be applied to both continuous process manufacturing as well as semicontinuous (batch) manufacturing.

The continuous process system is based on scheduling the entire process as a single unit. Finite forward scheduling is utilized to meet the desired requirements of the production plan. A simulation process will determine the best schedule to meet customer orders or planned finished-goods inventory levels. Raw material availability is also verified by this process, but only after available capacity has been ascertained.

Semicontinuous batch processing is managed with operation-by-operation scheduling. As with process scheduling, the system is production-plan-driven toward finished-product inventory levels. The planning process may start with the desired inventory and schedule each operation based on reverse flow or by forward flow scheduling starting with raw material. Which strategy to use will be based on process considerations such as bottlenecks or material availability. Each operation will be finite scheduled to meet planned inventory levels before moving onto the next (forward) operation or to the previous (backward) operation. In either case, material availability must be checked. Lot sizes and run times for each operation will be based on the specific process technology. The planned inventory levels may be based on lot sizes, curing times, or differing run rates. A bottleneck operation in the middle of the process may call for reverse flow scheduling to the bottleneck and forward flow scheduling after the process.

MRP COMPARED TO PFS

The scheduling of steel strap as defined in Table 10-2 can be accomplished by either the MRP logic of gross-to-net requirement controls at each operation or PFS scheduling based on process flow and by each work center's desired inventory levels and lot size rules. Both systems may incorporate finite scheduling and operation sequencing techniques.

Table 10-3 illustrates the MRP approach to scheduling a 150,000-pound requirement for product A.

The slitting operation for product A, the splitting operation for component F, and the cold-rolling operation for component E are then scheduled to execute the MRP plan.

Table 10-3. MRP Planning

Part	Lot Size	Lead Time	Time Periods in Weeks		
A	Lot for lot	1	1	2	3
Master schedule					150,000
Planned order release				150,000	
F	Lot for lot	0			
Gross requirements				150,000	
Scheduled receipts					
Projected on hand				−150,000	
Net requirements				150,000	
Planned order receipt				150,000	
Planned order release				150,000	
E	120,000 Minimum	1			
Gross requirements				150,000	
Scheduled receipts					
Projected on hand			80,000	−70,000	
Net requirements				70,000	
Planned order receipt				120,000	
Planned order release			120,000		

Table 10-4. PFS Scheduling

1. Schedule the slitting line to produce 150,000 pounds of product A by the beginning of week 3.
2. Schedule the splitter to split 150,000 pounds of component F to meet the starting date of product A on the splitting line.
3. Schedule the rolling mill to cold-roll component E to meet the 150,000-pound requirement and the starting date of component F on the splitter. If the rolling requirement calculates to less than 120,000 pounds (the minimum run size), roll 120,000 pounds and hold the excess in work-in-process.

Table 10-4 illustrates the PFS approach calling for the same requirement of 150,000 pounds of product A. Each operation is scheduled based on reverse flow.

CASE STUDY

Problems

The projected monthly sales demand for Chemical "A" is as follows:

January	10,000 gallons	July	28,000 gallons
February	11,000 gallons	August	19,000 gallons
March	13,000 gallons	September	11,000 gallons
April	15,000 gallons	October	10,000 gallons
May	20,000 gallons	November	9,000 gallons
June	25,000 gallons	December	9,000 gallons

The maximum finished product to be inventoried is 20,000 gallons—the capacity of the holding tank.

1. Calculate a production plan for product A based on an empty storage tank at both the beginning and the end of the year and with minimum setups.
2. It is determined that the shelf life of product A is 1 month. Recalculate the production plan still planning an empty storage tank at both the beginning and end of the year, but with no more than 1 month's usage in storage.

Solutions

1. The production plan calling for nine setups is as follows:

	Demand	Production	Month-End Inventory	Approximate Schedule Time
Jan.	10,000 gals	20,000 gals	10,000	First of month
Feb.	11,000 gals	20,000 gals	19,000	End of month
Mar.	13,000 gals		6,000	
Apr.	15,000 gals	20,000 gals	11,000	Second week
May	20,000 gals	20,000 gals	11,000	Second or third week
June	25,000 gals	20,000 gals	6,000	Second week
July	28,000 gals	40,000 gals	18,000	First and fourth weeks
Aug.	19,000 gals	20,000 gals	19,000	Last of month
Sept.	11,000 gals		8,000	
Oct.	10,000 gals	20,000 gals	18,000	Third week
Nov.	9,000 gals		9,000	
Dec.	9,000 gals		0	

2. The adjusted production plan which calls for 13 setups is as follows:

	Demand	Production	Month-End Inventory	Approximate Schedule Time
Jan.	10,000 gals	10,000 gals	0	First of month
Feb.	11,000 gals	11,000 gals	0	First of month
Mar.	13,000 gals	13,000 gals	0	First of month
Apr.	15,000 gals	15,000 gals	0	First of month
May	20,000 gals	20,000 gals	0	First of month
June	25,000 gals	40,000 gals	15,000	First of month and last week
July	28,000 gals	20,000 gals	7,000	Third week
Aug.	19,000 gals	12,000 gals	0	Second week
Sept.	11,000 gals	11,000 gals	0	First of month
Oct.	10,000 gals	10,000 gals	0	First of month
Nov.	9,000 gals	9,000 gals	0	First of month
Dec.	9,000 gals	9,000 gals	0	First of month

QUIZ

1. Process flow scheduling is based on bill of material structuring.
   a. True
   b. False

2. MRP can be utilized in a continuous-process industry environment to
   a. generate shop orders
   b. reschedule production
   c. control raw material purchases
   d. expedite

3. Process manufacturing operations may consist of
   I. mixing
   II. forming
   III. separating
   IV. job lot machining

   a. I
   b. II and III
   c. I, II, and III
   d. All of the above

4. Fabrication and assembly operations may be accomplished with
   I. process manufacturing
   II. job-shop manufacturing
   III. repetitive manufacturing

a. I
b. I and III
c. II and III
d. All of the above

5. The bill of material in a process manufacturing environment will tend to be shallow.
   a. True
   b. False

6. Process manufacturing can be in a __________ mode.
   I. continuous
   II. batch

   a. I
   b. II
   c. I and II
   d. Neither I nor II

7. In process manufacturing, the flexibility of the MPS is dependent on
   I. available raw material
   II. capacity

   a. I
   b. II
   c. I and II
   d. Neither I nor II

8. In process manufacturing, if lot size constraints require batch controls for WIP, __________ is recommended.
   a. reorder points
   b. rough-cut capacity planning
   c. MRP
   d. infinite scheduling

9. With continuous processing, a high degree of WIP shop-floor control is required.
   a. True
   b. False

10. The goals of repetitive operations are the same as process manufacturing's goals.
    a. True
    b. False

BIBLIOGRAPHY

Finch, B. J. and Cox, J. F., *Planning and Control System Design: Principles and Cases for Process Manufacturers*. Falls Church, VA: American Production and Inventory Control Society, 1987.

Moran, J. W., *Please, no MRP! We're continuous flow manufacturers,* in American Production and Inventory Control Society 36th Annual International Conference Proceedings, 1993.

Taylor, S. G. and Bolander, S. F., *Process flow scheduling strategies,* in American Production and Inventory Control Society 37th Annual International Conference Proceedings, 1994.

Turek, R., Having trouble implementing MRPII? Maybe you're a process planning candidate, *APICS, The Performance Advantage,* October 1994.

11
Just-in-time Execution

Material requirements planning (MRP) had its beginnings in material control in job-shop manufacturing. The evolution of material resource planning (MRP II) has continued to address the needs of the job shop. The development of the Just-In-Time philosophy in many companies has caused a shift from job-shop manufacturing to repetitive (or semirepetitive) operations. Contrary to some beliefs, MRP II continues to play a major role in the planning of capacity and operations as well as execution support of Just-In-Time manufacturing.

The Just-In-Time (JIT) execution of a manufacturing plan is but a part of JIT philosophy. The use of JIT methods and procedures for shop-floor control are considered in the narrow definition of JIT, which calls for the movement of only the required material to be at the right place at the right time with each operation closely synchronized to the next. The broader definition of JIT is a philosophy for guiding manufacturing management. This philosophy refers to the management of all manufacturing functions and their related activities. Related activities include marketing, product design, sales distribution, and finance. The goal of JIT philosophy is to achieve maximum customer service while improving both quality and productivity simultaneously. Emphasis is based on continuous improvement through the elimination of non-value-added activities.

Just-in-Time can also be defined as follows:

1. World-class manufacturing.
2. Value-added manufacturing.
3. Continuous improvement manufacturing.
4. Making your supplier carry your inventory. This may not be correct, but sorry to say it is often true.
5. The extreme use of common sense (a favorite of the author).

The challenge of controlling complicated manufacturing operations has been understood for many years, but the solutions did not seem achievable until the advent of third-generation computers in the 1960s. With the ability to store an almost unlimited amount of data and to compute at what seemed to be unbelievable speed, manufacturing systems were developed to control the many complicated manufacturing operations. The problem was that the systems, although sophisticated to a high degree, still did not always control the manufacturing operations as desired. The lessons learned from the Japanese have been to concentrate on streamlining the manufacturing operation itself and requiring less complicated manufacturing systems.

The essential ingredients of JIT have been defined as follows:

1. Total quality by continuously seeking to serve the customer better.
2. People involvement through participation of everyone at all levels of the operation.
3. JIT manufacturing methods emphasizing the continuous flow of materials through the manufacturing process.

This chapter concentrates on the understanding and implementation of JIT manufacturing methods.

PROCESS FLOW GOALS

Repetitive Manufacturing

The basic goal is to make the material flow in both job-shop and repetitive manufacturing operations as it flows in continuous process manufacturing. One of the challenges to achieving the goal is to convert job-shop manufacturing into repetitive manufacturing or to at least adopt repetitive manufacturing features. The nature of some make-to-stock environments such as television manufacturing call for repetitive production, whereas make-to-order or assemble-to-order environments call for nonrepetitive production at least at the end item level. Examination of the product structure of nonrepetitive end item products may indicate repetitive potential at the modular, subassembly, or option levels. Also to be considered is the degree of commonality of components within the total product line. Production planning should consider the repetitive features of those items below the end item level.

If the existing product structures show no repetitive features, effort should be made to simplify existing products as well as considering repetitive demand potential in new product design. An additional consideration should be the grouping (families) of end items or components based on similarity of manufacturing processes. Family scheduling will allow a higher degree of repetitive production.

Level Schedules

To plan for process flow, an initial requirement is a level schedule at the production planning or MPS level. The level schedule is a first step in

Table 11-1. Lot Size Effect on Inventory Levels

	Lot Size	
	200	20
Run frequency	Every 2 weeks	Daily
Demand variations between production	80–360	0–60
Safety stock required for deviations	160	40
Average lot size inventory (lot size/2)	100	10
Average inventory	260	50

achieving balanced manufacturing for the entire operation. Balancing the operation will be addressed later in this chapter. The level schedule, planned over a horizon of 6 months to a year, is implemented through a final assembly or end item schedule which is responding to actual demand. Although planning may be in weeks or months for raw material and component control, a flexible end item schedule should be controlled on a daily basis.

Because customer orders (i.e., sales demand) do not arrive in even daily requirements, the system must allow for a level final assembly schedule to meet the demands of the marketplace which are not level. In a make-to-stock environment, an uneven customer demand rate can be met through the stocking of finished goods. If the assembly (or end item) lot sizes are small, the required finished-goods stocking level is less and the degree of flexibility is greater.

Table 11-1 illustrates the difference in inventory levels of a make-to-stock item when the lot size is reduced from 2 weeks demand to daily production. The forecasted weekly usage is 100 per week with possible biweekly demand variations of 80–360 and possible daily variations of 0–60. The level of inventory is substantially reduced, schedule flexibility is increased, and customer service remains the same.

In make or assemble-to-order environments, there are no finished-goods inventories to compensate for unlevel demands. End item scheduling will be based on order backlogs rather than planned inventory levels. The larger the backlog, the more orders available to level the schedule with a favorable product mix. The negative of a large backlog is that the larger the backlog, the longer the lead time of delivery to the customer. The success of achieving a level schedule in make or assemble-to-order environments is dependent on the following:

- Raw material and component availability through product structuring and forecasting techniques

- Reduced manufacturing or assembly lead times
- Reduced nonmanufacturing lead times such as order entry, product design, and planning
- Quoted lead times that are acceptable to the customer, but are greater than the manufacturing cycle times

One JIT manufacturer was able to accept an order, schedule it, and manufacture it within 3 days. Competition was quoting 4–6 weeks delivery, so the manufacturer quoted 3–4 weeks delivery and therefore had sufficient backlog for level scheduling with a balanced product mix.

Design Simplification

The concept of designing for manufacturability is based on a simple design being easier to manufacture and that the fewer items to produce, the less complications with which to deal. Product rationalization is the design process that reduces the products, options, and complexities of manufacturing. An example of product rationalization is automobile manufacturers offering 4 or 5 option packages that may include 15 or 20 options rather than a list of 50 or 60 options.

Modular designs not only assist in forecasting and master scheduling, but in reducing manufacturing complexities through the standardized process routing of the modules. Various combinations of modules will allow for a relatively large number of end products, but with reduced total manufacturing processes.

When the process has been simplified, the levels in the product structure can be reduced. Including the subassembly operation into the assembly operation will not only assist in process flow but will eliminate an inventory level and the required transactions. This can be accomplished by "collapsing the bill", that is, removing the subassembly or coding the subassembly as a phantom which will allow the subassembly requirement to blow-through the structure and not be inventoried.

The total parts count and process routing of a production operation can often be reduced by the group technology (GT) philosophy. It is based on coding each part with a classification code that designates common functions, characteristics, and process routings. Formal GT programs have identified high numbers of both parts and routing proliferations and have been able to reduce part counts up to 80% and to standardize and rationalize many routings.

With reduced bill of material levels and simplified process routings, there are now programs underway to integrate bills of material and routing files. This would replace the existing cross-referencing routines utilized in chained-file management systems. Fully developed, this approach would be quite similar to the data base used in the process flow scheduling of process manufacturing.

Setup Reduction

In the past, a great deal of attention was paid to setup details such as: was setup labor direct or indirect or should the setup be made by the operator or a setup specialist or setup crew. What was not considered was an effort to reduce the setup. Treating the setup as fixed and then balancing the cost with carrying costs to determine the lot size created large work-in-process (WIP) inventories and long lead times.

Just-In-Time philosophy does not accept setups as fixed and is based on setup reduction processes. There have been substantial successes in setup and corresponding lot size reductions. The formal setup reduction process is based on group or team involvement, with the team consisting of shop-floor people with operators playing an important role. One of the pioneers in setup reduction was Shigeo Shingo of Toyota who developed the single-minute exchange of die (SMED) process. The basics of SMED have been the cornerstone of numerous programs that have resulted in dramatic setup reductions from days or hours to a few minutes. The successful results have been based on teamwork, an organized four-stage procedure, and common sense.

The first stage is to analyze the existing process and setup to determine if the setup is really required or if it can be eliminated through product grouping or process simplification. If the setup cannot be eliminated, the details of the machine, the tooling, the material, and the existing routines can be best captured by videotaping the entire operation.

The second stage is to identify which setup activities are internal and which are external to the operation. An external activity is that which can be performed while the machine is running. A simple example of an external activity would be bringing the next tool or die to the workplace location and preparing it in advance. Another example would be preheating a die before attaching it to a plastic-molding machine. An internal activity is that which can only be accomplished when the machine is stopped. Attaching a tool to a machine is an example of an internal activity. Once identified, the setup procedure should be reviewed to assure that the machine is running while external activities are taking place. This effort has historically reduced machine downtime, the real setup time, 30–50%.

The third stage of the process is the reexamination of internal activities to assure that they were properly identified and to then attempt to convert them to external setup. The preheating of the die for plastic molding mentioned above is an example of an activity that might have been originally heated as an internal activity, but then converted to external by preheating.

The fourth stage is the improving of both internal and external activities of the setup. Examples of improvement techniques are parallel operations, the use of functional clamps, and the elimination of adjustments. In situations involving large machines, parallel operations by two people may reduce the activity time by more than half due to the economies of movement gained by not having to move around the machine. An example of the use of functional clamps is the U-slot cut into a die so that a bolt is easily fit into the die and then secured with one turn of the nut. The use of shims or blocks in die setting can be used to standardize the height and eliminate adjustment time.

To view dramatic results of setup change planning and teamwork, watch the Indianapolis 500 on Memorial Day. A race car can get a complete change of tires, fuel fill up, and the windshield cleaned in less than 20 seconds.

Lead-Time Reduction

The lead time to fabricate an item or to assemble a group of items is the sum of lead times for each operation. The operation lead time consists of the following time elements:

- Queue
- Setup
- Run
- Wait
- Move

Of the five elements, the run time is the only time that value is being added to the product. The reduction or elimination on the non-value-added activities will reduce the lead time, reduce WIP, and assist in the goal of synchronizing the operations to allow for greater process flow.

If operations can be combined, all four non-value-added elements will all be reduced. Reduced lot sizes and level schedules will reduce queue time, which is usually the largest non-value-added element. Work stations adjacent to each other will eliminate wait and move time when one operation is complete and the material is immediately in queue for the next operation.

PLANT LAYOUT FOR FLOW

To achieve material flow in a manufacturing operation, attention must be paid to the physical aspects of production including plant layout, housekeeping, and visibility control features.

Cellular Manufacturing

Historically, plants have been organized by functional or clustered layouts such as the Saw Department, Milling Department, Grinding Department, and so forth. Flow patterns consistent with JIT methods are accomplished with cellular manufacturing which produces parts or families of parts in a single line of machines based on the product flow rather than functional grouping.

Cellular manufacturing requires the grouping of parts based not on their design but on process similarities, that is, routings that follow similar paths. Note that similar paths are required, not exact paths. The determination of parts with process similarities can be accomplished through group technology (GT), a parts classification system. For this reason, cellular layouts are sometimes referred to as GT cells. Companies without a formal parts classification system can determine process similarities through routing data review. Other requirements for successful cellular manufacturing are short setup times, a level of volume that allows daily operation, and multifunctional operators and supervision.

The output of the work cell is based on the number of assigned operators rather than machines. If the planned output rate is low, there will be fewer operators who are required to run more operations, thus the multifunctional requirement. Supervision must also be multifunctional in that each diverse machine function must be understood.

The flow advantages of cellular manufacturing are reduced lead times due to small lot sizes and reduced (or no) queues between operations. Lot size effect can be additionally reduced by overlapping operations and creating sublots. Teamwork within the cell is enhanced through visibility controls and communication. The close working environment of the team and immediate feedback have shown very favorable quality results. Minimal travel distances within cells have simplified material handling and reduced floor space needs. Organizing the work stations within a cell in the shape of a "U" further reduces space requirements, reduces operator walk time, and improves communications within the work team.

Organizing a work cell is no easy task. Problems can arise when a specific operation may call for a machine that will be utilized a relatively small percentage of the time. The problem may be solved by a part design or routing change or by routing certain parts in and out of the cell. In and out routings are often required for heat treating and painting operations. The most favorable manufacturing cell is when the product and volume allow the cell to be a dedicated line. The dedicated line is limited to a family of parts that run consistently with very little line balancing required.

An extension of the cellular manufacturing concept is the focused factory. The focused factory is organized to produce a product or product families that have overall process similarities. The focused factory may consist of an assembly line or lines being fed by manufacturing cells. The factory management organization has reduced complexity due to operating teams and a limited staff reporting to the factory manager. When one or more focused factories are organized within a larger manufacturing facility, it is called a "plant-within-a-plant."

In-Plant Storage

Material flow is improved through the concept of storing WIP parts at either the point-of-use (the inbound) location where the parts will be used or the outbound location after they are produced and waiting for a signal to be moved to the next operation. This concept eliminates the need for a stockroom and its attendant costs, reduces material handling, and allows for linear flow. To achieve this storage concept, the plant must be well organized and have a high degree of shop-floor discipline.

Standardized containers assist in controlled material flow through the manufacturing process. The containers are designed to allow for ease of movement by one person, requiring no more than normal lifting or moving activity. The number of parts per container will be based on the shape and weight of the part. The parts within the container are organized and stacked through the use of separators in the manner of egg cartons. Standardized and organized containers also assist in production counts and reducing part damage due to movement. Most companies that have analyzed container size requirements have found that one size of container will cover the storage needs of 80% of the parts.

Workplace Organization

One of the initial activities for JIT implementation is to review and assure that good housekeeping is a way of life in the plant. Too often in the past, a plant would be cleaned up in preparation for the annual physical inventory. Everyone would agree that it should stay that way, and 2 weeks later it would revert back to its normal disorderly way. JIT techniques require a well-organized, clean operating environment at all times.

Maintaining a proper workplace is the responsibility of the operators with assistance (not haranguing) from supervision. Efforts should be made to simplify the required items within the area such as materials and tooling. These items should have a fixed storage location, with those frequently used stored closest to the workplace operation. Constant discipline must be in place in order to assure continued cleanliness and order at each work station.

The workplace organization must allow for JIT visibility controls when scheduling operations within work cells or from one work cell or work center to another. Scheduling may also be controlled through schedule boards or screens that can be seen by everyone. Signal lights may send messages relating to required tool or material changes or to communicate machine malfunctions. A marked floor location for a specific part that is empty may be the scheduling message to produce that part.

SYNCHRONIZED SCHEDULES

Material flow is achieved through synchronizing each operation to the next operation. Synchronizing of operations requires a level balanced repetitive work load and a facility that has a layout organized for material flow.

MPS/MRP/FASControls

The MPS driving the MRP system in a JIT environment supplies planning information to both suppliers and the manufacturing facility. This information does not constitute the authority to ship purchased material nor to produce manufactured components. The final assembly schedule (FAS) is based on immediate customer orders or distribution (stocking) needs. These FAS requirements must be consistent with the MPS. The FAS (or end item schedule) listing specific units and their rate of production control the flow of material through the plant as well as material from the supplier. This control is called a pull system in that material is pulled from the supplying work center or supplier only when that material is needed.

Figure 11-1 illustrates the MPS/MRP/FAS relationship in a JIT pull system. The master schedule explosion (the MRP) supplies fabrication and supplier plans. The final assembly requirements are pulled from work centers D and E. D and E requirements are pulled from work centers A, B, and C. Purchased part components, and A, B, and C raw material requirements are pulled from the suppliers.

In contrast to Figure 11-1, Figure 11-2 (a repeat of Fig. 9-1) illustrates the conventional MRP push system used in job-shop environments. The material is pushed to the plant via purchase orders and pushed through the plant via shop orders. The purchase and shop orders are based on the assumption that the MPS and FAS requirements will be met as planned.

Mixed-Model Production

A level schedule at the end item level does not guarantee that all plant operations will operate with level repeating sequences. An automobile assembly line might operate with level schedules by a scheduling sequence of two-door cars one week and four-door cars the next week. However, the pull requirements for door fabrications would double every other week. The goal

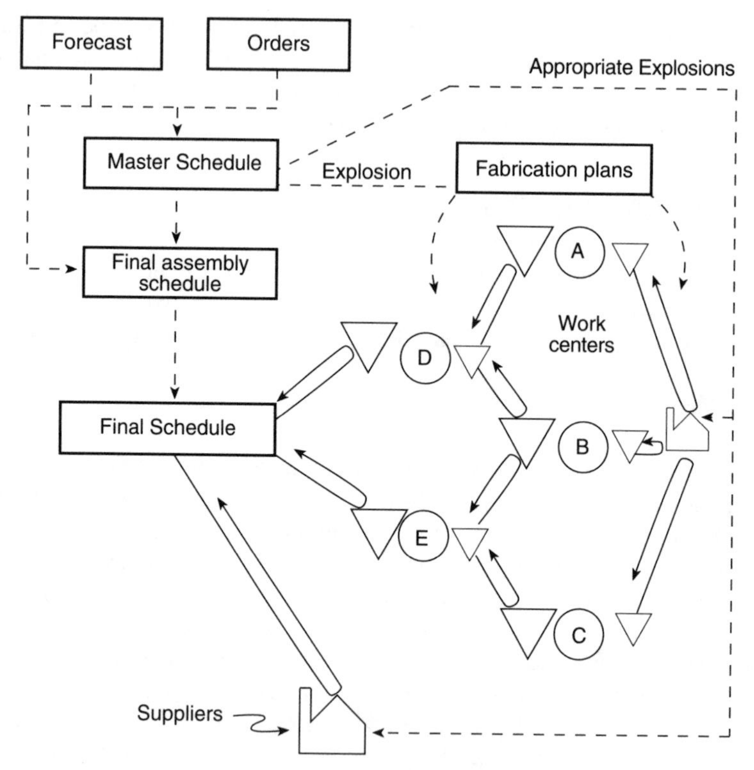

Figure 11-1. Demand-pull system. [Reprinted with the permission of APICS, Inc., Robert W. Hall, *The Implementation of Zero Inventory/Just-in-Time,* 1986.]

of uniform flow of materials through the entire operation is enhanced through the scheduling of mixed-model production at the end item level. With a desired lot size of one and a well-planned mixed-model end item schedule, the pull rate will allow synchronization of operations.

An example of mixed-model scheduling is illustrated in Table 11-2. A bicycle manufacturer has weekly assembly requirements of 600, 26-in.-frame, 300, 24-in.-frame, and 150, 20-in.-frame bicycles. The schedule pull rate or pulse should be uniform from bicycle assembly to frame subassembly, and frame subassembly to component parts.

Based on the above sequence and assuming equal assembly times of 2 minutes per bicycle, the usage rate of frames would be as follows:

26 in. four frames every 14 minutes
24 in. two frames every 14 minutes
20 in. one frame every 14 minutes

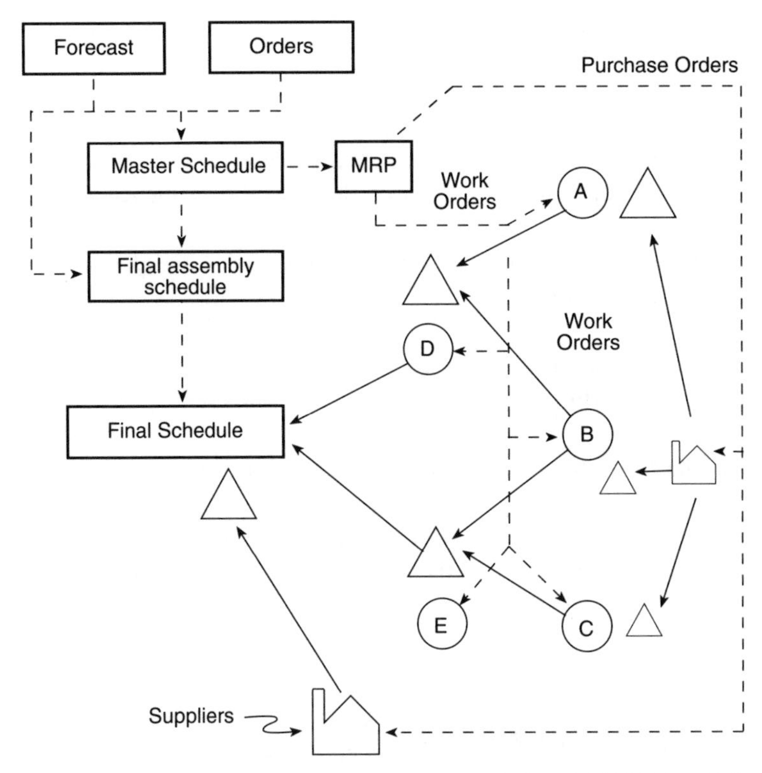

Figure 11-2. Order-push system. [Reprinted with the permission of APICS, Inc., Robert W. Hall, *The Implementation of Zero Inventory/Just-in-Time*, 1986.]

Table 11-2. Mixed Model Schedule

Model	Per Week	Per Day	Per Hour
26-in.	600	120	17.1
24-in.	300	60	8.6
20-in.	150	30	4.3

Note: The mixed model repeating sequence is 26-26-24-26-26-24-20.

Cycle Time

The cycle time is the time between completion of two discrete units of production. For example, if an item is produced at the rate of 90 per hour, the cycle time of the item would be 40 seconds.

The need cycle times for the frame requirements in Table 11-2 are as follows:

26 in. 14/4 = 3.5 minutes
24 in. 14/2 = 7 minutes
20 in. 14/1 = 14 minutes

Although the 26-in.-frame requirements will average one frame every 3.5 minutes, the actual usage rate based on the assembly sequence will be

one frame every 2 minutes for two assembly cycles and then
one frame for 4 minutes.

The 24-in.-requirement will average one frame every 7 minutes, but the actual usage rate will be

one frame for 6 minutes and then one frame for 8 minutes.

The difference in the uniform production cycle of the feeding work center and the uneven requirements of the pulling work center can be compensated by the pull system controls.

Pull Systems

The pull system in a JIT flow environment has two functions.

1. Bring the material to the using operation when it is called for
2. Authorize the replacement of the material

There are a variety of systems used for controlling the material flow and replacement. When there are a number of work centers and moves required in a production operation, a two-card system is used. The production card authorizes material replacement when it is removed from the container and is replaced with a move card. After completing production (the replacement material), the production card will be attached to the new container. The move card authorizes a move when it has been removed from the container at the using location and replaces a production card at the supplying location. Figure 11-3 details the flow path of cards at two work centers.

It can be seen that if an operation such as assembly stops, the movement of all cards through the supply line stops and production lacking the authority of production cards will cease. Move cards will have pertinent information such as part number and inbound and outbound location information. The production card information will include part number, container capacity, inbound and outbound locations, pertinent manufacturing details, and lot size

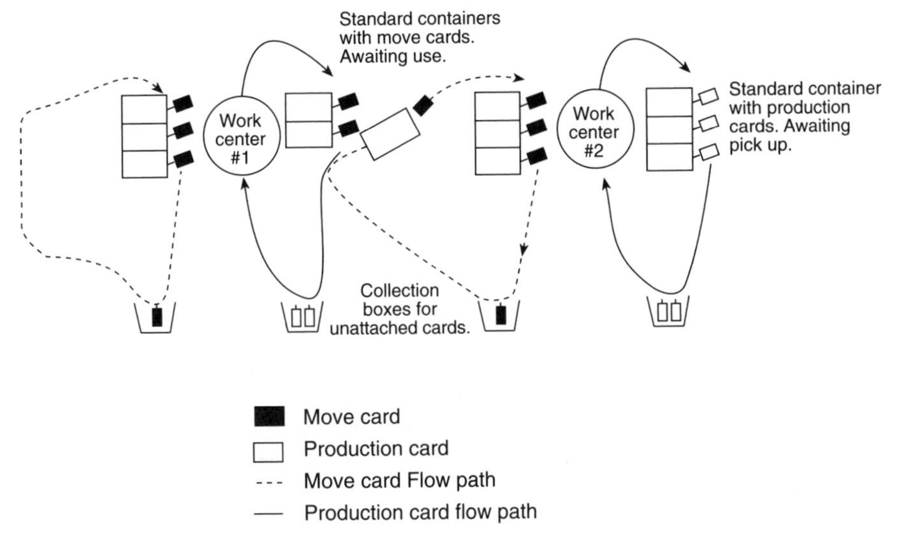

Figure 11-3. Detailed flow path of cards at two work centers. [Reprinted with the permission of APICS, Inc., Robert W. Hall, *The Implementation of Zero Inventory/Just-in-Time*, 1986.]

control such as produce in three-card quantities. In this case, production is not authorized until three cards are collected.

If the outbound location is adjacent to the supplying work center, a one-card system will work. An empty space in the designated outbound location is the authorization to replace an empty container with the required parts. The information normally found on a production card will be posted at the work center. The now full container will remain in the outbound location until moved via a move card generated by the inbound (using) work center.

Containers will serve as pull signals in less complicated flow operations. The container will be marked with a part number and specified quantity. When emptied by the using work center, it is returned for replacement at the supplying work center. Kanban squares will also serve as pull signals. There will be a designated space marked for specific materials. When that space is open (emptied), that is the authorization to replace the material. The two-bin system, which is older than JIT, can also be considered a pull system. When the first bin is emptied, that is the signal to replace a given quantity.

In the determination of the pull signal controls such as how many cards, containers, or location spaces are required, the expected usage rates, replacement lead times, lot sizes, and safety or buffer stocks must be determined. In the 26-in.-frame example where the replacement cycle was 3.5 minutes but the pull requirements of the assembly schedule was uneven, the difference could be covered by carrying an extra frame or two as buffer stock at the outbound location. This would be accomplished by an extra move card in the system.

SUPPLIER INVOLVEMENT

The reader will note that supplier involvement is the last subject discussed in this chapter. This is no accident. Too many companies have decided to become "JIT" companies and attempt to do so by the simple act of making their suppliers carry the inventory. This is wrong. In the first place, the goals of JIT are not achieved, and in the second place, if the suppliers are smart, they will eventually add the cost of carrying the inventory to the invoice. Worse yet, if they are not smart, they will either reduce quality or perhaps go out of business. Either way, the customer loses. Before expecting supplier involvement, the manufacturers should have their own houses in order relative to the JIT ingredients of total quality, people involvement, and JIT manufacturing methods. This is not to say that supplier relationships should be ignored. Supplier planning should fit into the overall initial JIT implementation program and become a part of continuous improvement efforts.

Extension of Manufacturing Operations

A successful approach to supplier involvement is to consider the supplier's operation as an extension of your manufacturing operation. The same process flow goals of small lot sizes and reduced lead times for manufacturing operations should be part of the supplier involvement program. Achievement of these goals will reduce the supplier's WIP and lead time, increase responsiveness to the customer, and reduce the customer's commitments. The success of these goals will be dependent on the communication ability of both parties. The supplier may be able to improve operations through simplification, material change, or specification revision if the end use of the product is understood. The customer may be able to make design changes to improve supplier manufacturing operations if the manufacturing processes are understood.

A simplified control system will save time and reduce costs. Efforts should be made to incorporate pull system controls between customers and suppliers similar to the pull systems used within manufacturing plants. A variety of pull signals have been successfully employed such as containers, cards, fax machines, or even telephone calls. Paperwork such as purchase orders, purchase releases, receiving reports, and invoices have been eliminated or reduced through back-flushing techniques. An example is to base the receipt, usage, and payment for tires on the number of automobiles assembled.

As with JIT implementation within the plant, the customer-supplier JIT program will require cross-functional understanding and education. The relationships must be understood, responsibilities defined, and specifications published that are realistic and doable.

Single-Source Suppliers

The advantages of single-source suppliers listed in Chapter 9 are especially true in JIT operations. The single-source supplier differs from a sole-source supplier in that there is a choice of going with a single source and, if so, which single source, whereas with a sole source, there is no choice. The single source may be a supplier of an entire product line or the single source may be stated at the part number level or be based on a specific location.

Implementation of the concept of the extension of supplier operations to manufacturing operations is more practical with the single-source approach. Product and process improvement programs are enhanced through the use of customer–supplier product teams. Communications relating to day-to-day questions and problems are much less complicated.

The JIT strategy of single sourcing has reduced the supplier base for many companies, allowing the purchasing function to spend more time on quality, delivery, and process improvement activities. The concerns of single sourcing such as paying too high a price or labor-related disruptions must be addressed, especially for critical parts. The Japanese companies that have taken the lead in many JIT techniques, including supplier involvement, will often establish a second-source supplier who may receive a reduced share of the business, but whose presence is available in emergencies as well as keeping the major supplier in line.

CASE STUDY

Problems

1. The output rate of a manufacturing cell feeding an assembly operation is controlled by the number of assigned operators. The variable piece rates are as follows:

No. of Operators	Pieces/hour
1	60
2	150
3	240
4	300
5	325

The assembly bill of material calls for two pieces per assembly and the assembly cycle time is 2.4 minutes. How many operators should be assigned to the cell for the most efficient operation?

2. A manufacturing cell feeds two assembly lines. Part A, which feeds assembly line 1, must be produced in lots of 500 units and takes 2 hours production time. Line 1 uses part A at the rate of 100/hour. The cell is adjacent to the assembly line and there is no appreciable move time, but a safety stock of 2 assembly hours is desired. The pull system is based on containers which hold 100 units each. How many containers are required?

Solutions

1. The cycle time of need is 2.4 minutes/2 = 1 piece every 1.2 minutes or a requirement of 60/1.2 = 50/hour. Therefore, one operator will allow for the most efficient JIT flow operation.

   Based on the traditional (and narrow) definition of an efficient operation looking at only direct labor, the most efficient operation would be based on three operators (80 pieces/hour per operator). This approach fails to consider the costs associated with extra inventory and interrupted material flow.
2. The container requirements are

Lot size of 500/100	5 containers
Assembly requirements during cell production run	
$2 \times 100/100$	2 containers
Safety stock $2 \times 100/100$	2 containers
Move time allowance	0
Total required	9 containers

QUIZ

1. Effective implementation of Just-In-Time should result in which of the following?
   I. Maximum work center utilization
   II. Employee satisfaction
   III. Development of detailed cost accounting data
   IV. Increased total business productivity

   a. I and III
   b. II and IV
   c. I, II, and IV
   d. All of the above

2. Long-term agreements with suppliers may include which of the following types of special arrangements?
   I. Simplified paperwork systems
   II. Returnable containers
   III. Specially designed containers
   IV. Effective personal contacts at each site

   a. IV
   b. II and III
   c. I, II, and III
   d. All of the above

3. To which of the following types of delivery can kanban methods be applied?
   I. Within a plant
   II. From plant to plant
   III. From suppliers to plants

   a. I
   b. I and II
   c. II and III
   d. All of the above

4. JIT training should increase the capabilities of operations supervisors to do all of the following EXCEPT
   a. balance work load in cells
   b. plan area production
   c. keep the work force running its machines
   d. coordinate preventive maintenance

5. Which of the following will occur when MRP and JIT are integrated?
   a. MRP will continue to support both operations planning and control.
   b. MRP will continue to support operations planning and JIT pull systems will support operations control.
   c. MRP will continue to support operations control and JIT pull systems will support operations planning.
   d. JIT pull systems will eliminate the need for MRP.

6. Small-lot production contributes to reduction of all of the following EXCEPT
   a. work-in-process inventory
   b. need for operator skills
   c. fluctuation of load on work centers
   d. the number of defective units

7. One of the objectives of JIT flow through production is to maximize machine utilization.
   a. True
   b. False

8. In the implementation of JIT, which of the following is the most significant change for the accounting function?
   a. Cellular processes no longer report every operation to accounting.
   b. Budgeted scrap and rework costs that create variances are reduced.
   c. Physical inventory of work-in-process inventory is more time-consuming.
   d. Costs of incentive programs increase.

9. Characteristics of parts containers appropriate for a JIT pull system include all of the following EXCEPT
   a. a design that allows their movement by hand without undue strain
   b. physical size appropriate for the consuming work center
   c. capacity appropriate for the lot size of the producing work center
   d. a means of protecting the parts from handling damage

10. Which of the following software capabilities is used so that transaction costs are minimized when JIT is implemented in an MRP environment?
    a. Pegging
    b. Back-flushing
    c. Forecasting
    d. Lot sizing

BIBLIOGRAPHY

APICS Dictionary, 7th ed., Falls Church, VA: American Production and Inventory Control System, 1992.

Hall, R. W., *Zero Inventories*. Homewood, IL: Dow-Jones Irwin, 1983.

Hall, R. W. and Ippolito, M. E., *Just-in-Time Certification Review Course*, Falls Church, VA: American Production and Inventory Control Society, 1992.

Hall, R. W., *The Implementation of Zero Inventory/Just-in-Time*, 1988.

Hastings, N. A. J. and Chung-Hsing Yeh, Bill of manufacture, *Production and Inventory Management Journal*, pp. 27–31 (4), 1992.

Jordan, H. H., Are you ready for Just-in-Time? APICS, The Performance Advantage, November 1994.

Shingo, S., *A Revolution in Manufacturing: The SMED System*. Stamford, CT: Productivity Press, 1983.

Toomey, J. W., Establishing inventory control options for Just-in-Time applications, *Production and Inventory Management Journal*, (4), 1989.

Womack, J. P., Jones, D. T., and Roos, D., *The Machine that Changed the World*. New York: Harper Collins Publishers, 1990.

12 MRP Implementation

When there is an understanding of what an MRP system should do and how it relates to the manufacturing operation, the final and perhaps most difficult, if not the most lengthy step, is the implementation of the system. Because the success of the MRP planning system must be measured by the execution of the plan, that plan must be compatible with the resource requirements of the manufacturing process. The manufacturing system requirements must be defined in order to design the proper system. Once designed, the costs of implementation must be understood and justified.

The establishment of performance measurements prior to implementation are required not only for justification but for continual monitoring of the new system once installed.

Project management must be planned to include education, project team organization, identification and resolution of issues, and a software selection process. Finally, the implementation program must include employee training and cut-over scheduling.

SYSTEM REQUIREMENTS

Requirement Defined

It has been said that properly defining a problem is one-half of the solution. It is better to consider system requirements to be the listing of opportunities rather than problems. Material requirements planning (MRP) should not be considered just a production and inventory control system, but a manufacturing plan with specific goals. There must not only be an understanding of the desired manufacturing system but also a full awareness of its relationship to other functions such as sales, distribution, and finance.

The amount of effort required for implementation will be dependent on the starting point. If the program is one of reimplementing an existing system, many of the basic requirements may be in place. If starting from ground zero, plans must include the implementation of basic data such as bill of material, routing, and item master files. In addition to data information, there must be the discipline in place to assure data integrity such as accurate bills of material and inventory records.

Specific requirements will depend on the nature of the organization and the manufacturing environment. A multiplant operation may call for a single master production schedule driving multiple interrelated MRP systems. A manufacturing system supplying complex distribution operations may require a simultaneous DRP implementation in order to properly control the master production schedule.

A job-shop operation will probably require capacity requirements planning at the work center level and shop-floor control dispatching by operations. Consideration must be given to prioritizing by either finite or infinite loading. A Just-In-Time (JIT) flow shop operation will call for material and capacity planning, but with simplified execution needs. JIT operations without work orders and work center schedules will require minimal shop-floor controls. A continuous process flow operation may require detailed capacity planning, but the shop floor may be directly controlled by customer orders or the master production schedule (MPS) rather than MRP. There may be special requirements for certain industries such as traceability for pharmaceuticals, accounting requirements for government orders, and shelf life for certain products.

The System Design

Once the scope and the basic requirements of the system are defined, the system must be designed. The following subsystems and their related features are design considerations. The subsystems address the MRP II activities shown in Figure 12-1 (a repeat of Fig. 1-3).

Business and production plans are developed based on historical data as well as sales forecasts. The master schedule may consider existing inventories, desired inventory levels, customers orders, load leveling, and capacity constraints. The MPS may be based on either end items or planning bills. Customer order processing considerations may include electronic data interchange (EDI) systems, product pricing and estimating, order entry routines, shipping, and sales history.

Both the product and process must be defined. The product structure may require definition by end items, modules, and/or options. An item master listing every part and the attributes of each part is necessary. A process routing (and possibly alternate routings) will be needed, with the level of detail dependent on the manufacturing environment. Work center data required will, like routing data, be dependent on the manufacturing environment. The product and process data may also be utilized for standard costing activities. Engineering change order procedures will be a consideration.

Material planning is based on MRP logic of net requirements determination

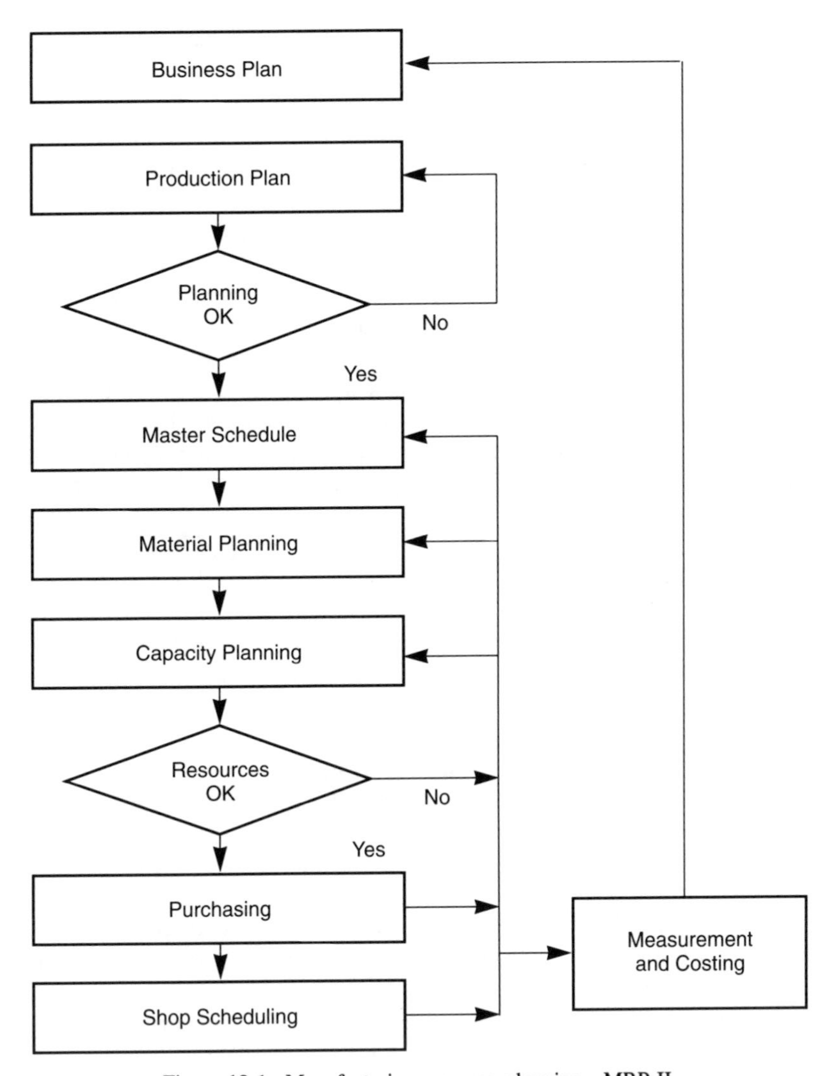

Figure 12-1. Manufacturing resource planning—MRP II.

and lead-time offsetting. Examples of the attributes for MRP systems to be considered are lot-sizing techniques, firming planned orders, and pegging. The nature of the operation will determine the frequency of the planning cycle. Planned purchase order and work order release routines must be defined. Inventory control systems must include inventory on hand and open work and purchase orders. They must also include both projected inventory on hand and scheduled receipts by time periods. Part allocation routines most often will be required. Reorder point control may be necessary for items coded as non-MRP.

Resource planning considerations should consider key resources and bottleneck constraints to allow for rough-cut capacity planning at the production plan and MPS levels. Capacity planning at the MRP level requires work center operating details such as hours available, efficiency, and utilization factors. The level of scheduling and dispatching of operations is dependent on the manufacturing process. A decision must be made whether to use infinite loading or finite loading and what prioritizing rules will be utilized. Operation sequencing may also be a consideration. Shop-floor reporting requirements must be determined and will be based on scheduling and dispatching systems, as well as required performance measurements. Job costing may be necessary for work-in-process (WIP) inventory valuation and labor and material variance calculations. The scope of the system may also include financial management features such as general ledger, accounts payable and receivable, and asset management.

System Justification

Project justification is normally based on a cost–benefit analysis and can be a most difficult task. The cost of MRP implementation can be reasonably estimated and should include the following:

1. The cost of both full-time and part-time personnel whose activity is strictly devoted to the implementation. Management information system (MIS) personnel should only be included if their presence is due only to the program and they would not be on the payroll for any other reason.
2. The cost of additional hardware required for the system.
3. The cost of purchased software, including implementation and contract programming.
4. Related outside expenses such as education and consulting.

The anticipated benefits are much more difficult to project. Benefits expected from MRP implementation are as follows:

1. Increased sales due to improved customer service. This is a real benefit, but the determination and auditing of a sales increase due to improved service can be most difficult. A 10% sales increase may be partially due to improved service, but market conditions, advertising, and quality improvements may have also contributed. In a market downturn, sales may drop 10% when improved customer service limited the drop to 10% rather than 15%.

2. Inventory reduction. Many MRP installations have shown dramatic improvements due to inventory reductions. Based on turnover analysis and cost of money estimates, reliable savings can be estimated and monitored. These calculated benefits alone sometimes are great enough to justify the system implementation. Although this is favorable from a justification viewpoint, the problem is that inventory reduction is not the reason for MRP, but is a favorable by-product.
3. Other benefits attributed to MRP implementation are increased productivity due to less part shortages and reduced expediting, reduced obsolescence due to forward visibility, and improved quality due to schedule stability. The capability to deexpedite assists in inventory reduction and work-load control. As with increased sales, these benefits may be real but quite difficult to calculate due to other variables. Estimated benefits that cannot be measured are of questionable value.

The application to the company's cost justification format of anticipated implementation and operating costs, inventory reduction cost benefits as well as very conservative estimates of sales, productivity, obsolescence, and quality benefits may well justify the program. If so, the numbers should be also utilized for ongoing audits of the system implementation and operation.

If unable to justify the program on the above cost–benefit basis, justification may be based on less tangible considerations such as the realization that competition is gaining in market share, at your expense, due to superior customer service. Another consideration may be that the existing system is adequate, but anticipated growth, product changes, or process innovations may call for a more formal and/or sophisticated system in the foreseeable future. A dollar benefits estimate, although soft, can be projected based on what is forecasted in the future. This benefits estimate can be compared to the anticipated costs.

However the implementation is justified, the calculated cost figures can then be utilized for cost control of the implementation.

PERFORMANCE MEASUREMENTS

With the implementation of a new system, measurements should be in place to measure the performance of that system. Measuring performance is fundamental to accountability and the purpose should be to stimulate improvements. Measurements may be used to establish goals (benchmarking) which will assist in achieving manufacturing excellence. Two broad categories for performance evaluation are financial and operational. A profit and loss statement is an example of a financial measurement, whereas customer service on-time delivery is an operational measurement. Inventory turnover rates, although calculated with dollars, are measurements of the material and manufacturing operation.

Performance measurements may or may not be easily related to the MRP implementation. Inventory accuracy measurement is required to make the MRP system reliable, whereas schedule performance evaluates how well the system is operating. Quality performance may be favorably impacted by a well-run MRP system, but there are many other factors affecting quality. The planned measurements calculated in the MRP justification should be measured not only after implementation but before implementation, so as to compare and justify the program. A 98% customer service measurement after implementation is meaningless if the service level prior to implementation is unknown (maybe it was 99%).

RULE OF MEASUREMENT

There are requirements and actions necessary for the establishment of performance measurements. Accurate and available operating data that will have an impact is a requirement. Performance goals such as 99% inventory accuracy should be established. Some measurements such as sales dollar per employee may not lend themselves to goals, but the measurement system should be used for trend review.

Measurement details that must be determined are as follows:

1. Type of Measurement—An example would be a measurement of inventory investment.
2. The Level of Measurement—Inventory investments might be a global measurement (total dollar turnover rate) or a more detailed measurement such as days of raw material on hand.
3. The Time Period of Measurement—The total inventory dollar turnover rate might be calculated monthly, whereas the days of raw material on hand are calculated every week.
4. Measurement Formula—An example would be the inventory turnover rate which is cost of sales/inventory investment.
5. Tolerances—Some measurements require tolerances such as a cycle count that is ± 0.5% of the stock status record and is considered accurate or a customer order that ships within 2 days of requested ship date and is considered on time.
6. The Goal of the Measurement—The customer service goal of on-time delivery might be 100% or the inventory accuracy goal might be 98%.

Company Performance Measurements

Table 12-1 is a listing of measurements which can be indirectly affected by an MRP system but are also affected by other influences. The measurements

Table 12-1. Company Performance Measurements

Measurements	Description
Return on investment (ROI)	Net income related to total assets and expressed as a percentage
Customer service	Orders or units shipped by promised date
Sales performance	Actual sales compared to plan or forecast
Quality improvement	Defects and quality costs compared to improvement goals
Cost variances	Actual labor, material, and overhead costs compared to standard costs
Lead-time reduction	Supplier or shop lead times compared to reduction goals

are examples of performance evaluations. Each company's measurements should be tailored to the operating requirements.

MRP Performance Measurement

Measurements of activities required for the successful implementation and ongoing operation of an MRP system are shown in Table 12-2.

As with companywide measurements, there are variations of the listed measurements as well as additional measurements that might be appropriate such as work center outputs or pick list fill rates. The detailed list should be based on the specific need of the manufacturing operation.

Table 12-2. MRP Performance Measurements

Measurement	Description
Inventory accuracy	Cycle count comparison to stock status record
Bill of material accuracy	Comparison of what is required to build an item to the computer bill of material record
Routing accuracy	Comparison of actual process sequences, operations, and work centers to the routing file
Delivery performance	Both supplier and shop conformance to scheduled receipt
Master production schedule performance	On-time completion compared to schedule
Inventory investment	Inventory turnover based on relating inventory to actual or projected cost of sales
System stability	Number of rescheduling and expediting actions required compared to stability goals

Rating of the MRP System

Table 12-3 is an example of rating classifications and their related characteristics. The performance percentages are based on the grading of selected (by the user) performance measures.

Table 12-3. Class A, B, C, and D Descriptions

Classification	Performance	Characteristics
A	90%	Complete closed-loop system. Top management uses the formal system to run the business. All elements average 80–100%.
B	80%	Formal system in place but all elements are not working effectively. Top management approves but does not participate. Elements average 80–90%.
C	70%	MRP is order launching rather than planning priorities. Formal and informal system elements are not tied together. Some subsystems are not in place. Elements average 65–75%.
D	50%	Formal system not working, or not in place. Poor data integrity. Little management involvement. Little user confidence in system. Elements are 50% or below.

Source: Reprinted with the permission of APICS, Inc., D. Buker "Performance Measurement," American Production and Inventory Control Society 23rd Annual International Conference Proceedings, 1980.

PROJECT MANAGEMENT

Organization

The MRP implementation effort requires guidance and control at multiple levels within the organization. The executive steering committee should be made up of top management people and take an active role meeting at least once a month. The committee's responsibility will be to review the implementation status, guide the project team, and make resource decisions when required. Monitoring of implementation costs is important especially if modification becomes extensive. The project team's responsibility will be the following:

1. Establish a project schedule
2. Report performance to schedule
3. Identify problems
4. Establish priorities
5. Reallocate resources
6. Establish formalized production goals

The project team should meet weekly to review the activities of the various task forces. Task forces are spin-offs of the project team with responsibilities for specific tasks such as the implementation of the MPS module.

The project team leader should have an operations background in the company, have the respect of the people, and be the best available person. If the potential leader does not have materials experience, MRP is more easily taught than having to learn the operations (and people) in the company. If at all possible, the project leader should be freed from other responsibilities and serve full time on the project implementation. Other members of the team can serve on a part-time basis, but the amount of time should be defined and committed to at the beginning of the project. Recommended team members should be at the manager or supervisory level and represent sales, customer service, accounting, production, production control, purchasing, distribution, human resources, engineering, quality control, and data processing. The amount of commitment time will be dependent on the function.

The use of an outside consultant should be limited to the following:

1. Assisting the team with technical or educational problems
2. Helping to focus on proper priorities
3. Pointing out potential implementation problems or roadblocks to the proper authorities
4. Assisting in solving issues in the most cost-effective way

Education

The education goal for implementation is to teach the users the "why" of the MRP system. The "how" to operate the system will be part of the training activity which is associated with the software implementation and will follow education. The education should cover the MRP body of knowledge which has been addressed in this book. Concepts of MRP must be first understood and then acted upon. The successful education program will not only teach the principles, concepts, and techniques, but that knowledge will also reduce resistance to change which can happen when change is not understood.

The basic requirements for accomplishing the MRP education are the following:

1. Active Top Management Participation—To properly guide the implementation, top management (the steering committee) must understand the basics. Top management participation also sends a message of the importance of the program.
2. Accountability by Key Operating People—MRP is a manufacturing system, not a data processing or production control system. Membership on the project team and the related education is a reflection of key operating users' involvement.
3. Education for Everyone—The only people who need to understand MRP are the people who use it. Who is going to use and be affected by

MRP? Everyone. The level of education will depend on the employee's function.

4. A Good Educational Program—The program must be well designed and taught by competent instructors to properly organized classes. The organization should allow uninterrupted classes.

The course content will be dependent on the participants. There should be programs tailored to the following:

1. Top management—The Steering Committee
2. The project team
3. Middle management
4. First-line supervision and staff
5. Operating people—The Production Workers

The decision of the use of education resources such as outside executive seminars, in-house education, consultant teachers, and training the trainers will be dependent on the environment of the company.

Software Considerations

In the early days of basic MRP implementation, many companies wrote their own software. There were successful implementations, but many found that this approach took longer than anticipated, was hard to get the software up and running, and, once running, found it to be inflexible relative to required modifications. Because the body of MRP knowledge has increased, so has the capability of commercial software packages. This is not to say that there are no problems with software that are too complicated, not complete, or difficult to interface. The key is for the user to do an intelligent job of software selection.

A starting point in software selection might be a commercially available software selection checklist. The list may number 500 or more specific MRP features for consideration. Each feature should be evaluated and determined to be either required, desired, or not needed. This evaluation routine will also assist in defining the needs of the system. The completed checklist of requirements and desires can be submitted to software manufactures for their capability responses to the checklist. A disadvantage of this response procedure is that the software manufacture is limited in stressing specific features that may be critical.

The software selection process must consider the following:

1. Are you really ready to make the software decision? In other words, are you comfortable with your system plans?

2. Does the software offer options and features that will meet all the known needs but in doing so be quite complicated and require a very high degree of user software knowledge? The alternate approach is to purchase a less complicated (slightly vanilla) system which can be modified to meet the user's needs.
3. Does the software have a good track record? This can be determined through contact with companies using the software or with user groups. As attractive as new software with new features may sound, remember that those on the cutting edge may bleed profusely.

Once the software has been selected, plans for the implementation must be made. Bridge programs that will transfer data from the old system to the new system must be written. Interfacing existing application programs to the new software must be planned. Finally, getting the new software to run on the computer must be accomplished.

ACTIVATING THE SYSTEM

The first step in software activation is to get the software up and running on the computer. This will be the responsibility of the MIS systems and data processing people. A test program supplied by the software manufacturer will assist in this task. The bridge programs written to transfer data from the old system to the new should also be tested at this time.

TRAINING

Once the software is in place, the next step is to educate and train the users to understand and use the software relative to their functions and responsibilities. This training process is known as the conference room pilot. Representative samples of company data are used in this procedure, but live data are not to be used. The objectives of the pilot are the following:

1. Train the people to use the required system operating procedures such as data entry, transaction reporting, order releasing, and so forth.
2. The users must understand the feedback from the system and the proper reactions to the feedback. An example would be an exception message recommending a change in the MPS schedule.
3. Training in the effects of a response such as changing the MPS schedule as recommended above. The planner making the change should understand the schedule changes effect on customer service and manufacturing resources.

4. The training process in the conference room pilot will also detect system bugs or shortcomings prior to live system activation.

Cut Over

Going to a live pilot is the application of the system to one group or family from the total operation. The pilot group is removed from the old system and run on the new. Considerations in the determination of the pilot group are as follows:

- It should be a representative sample of the business.
- It should be all items within a product family or a production line.
- It should be as self-contained as possible, in that there are a minimum number of parts common to both the pilot group and the rest of the business.

The pilot group should run alone until the users are satisfied and comfortable with the system. The run period should at least cover the time period of the manufacturing planning cycle.

When the initial pilot group is up and running successfully, additional groups may be added one at a time or all at once. One group at a time is the more conservative approach in that it is less risky and the user training period will be in series and less intense. Conversion of all remaining groups at once has the advantages of all common parts running on one system, dual inventory systems eliminated, and all plant and purchasing requirements on the same system.

There will be cut-over problems that must be understood and planned for. There may be a temporary inventory buildup due to the new system causing the move in (expediting) of needed requirements while the old system was unable to move back (deexpedite) previous orders which are now in stock and not needed. Due to tighter controls, the new system may call for temporary reduction of production and purchase order releases. In order to maintain a reasonable work flow, early releasing might be needed. Care also must be taken with timing to assure that the stock status updating of the new system is synchronized to MRP planning calculations. If part allocations are not updated prior to the MRP run, the system will fail to recognize certain needs.

Monitoring

Auditing is required not only of the system's performance but also of the implementation process itself. Continual attention must be paid to the following:

1. Are all the assigned implementation tasks on schedule?
2. Are all questions from all levels being answered and are problems being addressed?
3. Is the education program underway as planned?
4. Are the users being properly trained for system operation?

In addition to implementation monitoring, an audit of the system's capabilities is required in the following areas:

1. Is there sufficient computer memory?
2. Is there sufficient peripheral equipment such as terminals and printers?
3. Are the transactions handled in a timely fashion?
4. Are program generations and response times in line with original estimates?

Key performance measurements should have been in place prior to implementation, especially those measurements used for justifying the new system. Inventory, bill of material, and routing accuracy measurements should not only have been in place but their measurement goals met. The best software and most highly trained users will not be able to make the MRP work well if the data are not accurate. MRP performance measurements similar to those listed in Table 12-2 and company performance measurements similar to those in Table 12-1 should be implemented in order to meet the grade of a "class A" user.

CASE STUDY

Problems

1. An MRP implementation has not gone as well as expected. Which one of the following might have been the detriment to the implementation of the MRP system?
   a. The software was not picked out until the process is defined.
   b. The first group to receive MRP education was the top management.
   c. The project leader was a career manager from data processing.
   d. Performance measurements were taken prior to implementation.
2. Prior to implementation of an MRP system, the following key measurements were recorded. What should be the priorities to achieving the measurement goals.

	Actual	Goal
Customer service on-time delivery	60%	98%
Inventory accuracy	70%	98%
Inventory turns	3.1	5.0

Solutions

1. The detriment was picking the project leader from data processing. The project leader should have been a person with operating experience who will approach implementation from a user's viewpoint. The other three factors were exactly as they should have been.
2. The first priority is to reach the inventory accuracy goal. The MRP system will not work without it. The second priority is to achieve the customer service level. Based on the existing systems performance, it will require the MRP to achieve the 98% goal. Improved inventory turns may take a longer time due to excess inventory left over from the old system.

QUIZ

1. In the implementation of MRP, the leadership role should be
   a. production control
   b. top management
   c. data processing

2. In MRP implementation, which of the following is an initial step?
   I. Project organization
   II. Education
   III. Performance goals
   IV. Software selection

   a. I
   b. I and III
   c. II
   d. IV

3. MRP education is required at what organization levels?
   I. Top management
   II. Manager
   III. Supervisory and staff
   IV. Operator

   a. I and II
   b. II
   c. II and III
   d. All of the above

4. MRP implementation is the primary responsibility of operating people rather than data processing.
   a. True
   b. False

5. Specific detailed implementation issues should be addressed by
   a. the steering committee
   b. the project team
   c. the task force

6. The major system training emphasis should occur during
   a. justification
   b. initial MRP education
   c. the conference room pilot
   d. software selection

7. Performance measurements should be initiated
   a. prior to MRP implementation
   b. when the conference room pilot is underway
   c. during the initial cutover
   d. when implementation is complete

8. Data accuracy goals can be more easily attained once the MRP system is successfully implemented.
   a. True
   b. False

9. Bridge programs are required for:
   a. data accuracy
   b. software evaluation
   c. companywide goal establishment and monitoring
   d. data transfer from the old to new system

10. During the initial implementation of a new MRP system, inventory levels may:
   a. increase
   b. decrease
   c. remain the same

BIBLIOGRAPHY

Buker, D., *Performance Measurement,* American Production and Inventory Control Society 23rd Annual International Conference Proceedings, 1980.

Fogarty, D. W., Blackstone, J. H. Jr., and Hoffman, T. R., *Production and Inventory Management*. Cincinnati: South-Western Publishing, 1991.

St. John, R. E., *Material and Capacity Requirements Planning Certification Review Course*. Falls Church, VA: American Production and Inventory Control Society, 1991.

Toomey, J. W., Adjusting cost management system to lean manufacturing environments, *Production and Inventory Management Journal* 35(3), 1994.

Wallace, T. F., *MRP II: Making it Happen*. Essex Junction, VT: Oliver Wight Limited Publications, 1990.

Wight, Oliver W., *MRP II: Unlocking America's Productivity Potential,* Williston, VT: Oliver Wight Limited Publications, 1981.

APPENDIX A
Glossary

Many of the terms are adapted from the *APICS Dictionary,* (7th ed., 1992). It is reprinted with permission from the American Production and Inventory Control Society, Inc.

Allocation An allocated item is one for which a picking order has been released to the stockroom but has not yet been sent out of the stockroom. It is an "uncashed" stockroom requisition.

Assembly A group of parts and/or subassemblies that are put together and constitute an end item or a major subdivision for a final product.

Assembly Order A manufacturing order to put components together into an assembly.

ATP Available-to-Promise.

Available-to-Promise (ATP) The uncommitted portion of a company's inventory and planned production, maintained in the master schedule to support customer order promising.

Bill of Material (BOM) A listing of all subassemblies, parts, and raw materials that go into a parent assembly showing the quantity of each required to make the assembly.

Bill of Material Processor A computer program for maintaining and retrieving bill of material information.

Bill of Resources A listing of the required capacity and key resources to manufacture a given item or family. The resources are further defined by a lead-time offset.

BOM Bill of material.

Bottleneck A facility, function, department, or resource whose capacity is equal to or less than the demand placed upon it.

Business Plan A statement of long-range strategy of revenue, cost, and profit objectives. It is usually stated in terms of dollars and grouped by product family.

Capacity Requirements Planning (CRP) The process of determining in detail how much labor and machine resources are required to accomplish the open shop orders and planned orders of an MRP generation.

Carrying Costs Costs of carrying inventory usually defined as a percentage of the value of the inventory and includes the cost of capital invested as well as the cost of maintaining the inventory, such as taxes, insurance, obsolescence, space, and manpower.

Cellular Manufacturing A manufacturing process that produces families of parts within a single line or a cell of machines operated by machinists who work within the line or cell.

Central Supply Center The stocking facility with finished goods and/or service items and which supplies and maintains a distribution system.

Closed-Loop MRP A system built around material requirements planning that includes production planning, master scheduling, capacity planning, and the means for executing the capacity plans, material plans, and vendor schedules. The term "closed loop" implies that there is feedback provided by the execution functions so as to maintain a valid plan.

Component Raw material, part, or subassembly that goes into a higher-level assembly.

Conference Room Pilot An implementation process where users are trained to use the new software and identify potential problems. Dummy rather than live data are used.

Continuous Improvement A never-ending effort to expose and eliminate root causes of problems.

Continuous Processing Lotless production in which products flow continuously rather than being divided.

CRP Capacity Requirements Planning.

Cumulative Lead Time The longest planned length of time involved to accomplish a given activity. For an assembly, it would be that material path with the longest accumulated times.

Cycle Time In Just-In-Time planning it is the time between completion of two discrete units of production, such as motors assembled at the rate of 120 per hour would have a cycle time of 30 seconds.

Deexpedite The reprioritizing of jobs to a lower level of activity. Moving the due date back on the schedule.

Demand Filter The monitor of actual data for individual items in forecasting models. It is tripped when a demand is beyond reasonable deviation.

Dependent Demand Demand directly related to or derived from the bill of material structure for other items or end products and are, therefore, calculated and should not be forecasted.

Dispatch List A listing of manufacturing orders in priority sequence. Dispatch lists are normally generated daily and by work center.

Distribution Center A warehouse with finished goods located to be close to customers.

Distribution Inventory Finished-goods inventory to meet customer demand as opposed to manufacturing inventory to meet production requirements.

Distribution Requirements Planning The time-phased, net requirements explosion of warehouse needs via MRP logic.

Distribution Resource Planning (DRP) An extension of distribution requirements planning and considers the resources of the distribution system.

DRP Distribution Resource Planning.

Economic Order Quantity (EOQ) The mathematical computation for determining the amount of an item to be manufactured or purchased and is based on when the ordering cost of the lot size is equal to the carrying cost.

EDI Electronic Data Interchange.

Electronic Data Interchange The paperless electronic exchange of trading documents, such as purchase orders, shipment authorizations, advanced shipment notices, and invoices.

End Item A product sold as a completed item or repair part.

EOQ Economic Order Quantity.

Execute To carry into effect the manufacturing plan.

Expedite To rush production or purchase orders that are needed in less than normal lead time. Moving the due date up on the schedule.

Explosion The process of calculating the demand for the components of a parent item by multiplying the parent item requirements by the component usage quantity specified in the bill of material.

Exponential Smoothing A type of weighted moving-average forecasting technique in which past observations are geometrically discounted according to their age. A smoothing constant is applied to the difference between the most recent forecast and the latest actual sales.

FAS Final Assembly Schedule

Final Assembly Schedule (FAS) A schedule of end items to complete the product for customer orders and/or distribution inventory.

Finished Goods A product sold as a completed item or repair part.

Finite Loading Assigning no more work to a work center than the work center can be expected to execute. The specific term usually refers to a computer technique that involves calculating shop priority revisions in order to level load operation by operation.

Firm Planned Order (FPO) A planned order frozen in both quantity and time in order to assist the planner in responding to material and capacity problems.

Fixed Order Quantity A lot-sizing technique based on a predetermined fixed quantity and in which the frequency of ordering may vary.

Fixed Period Quantity A lot-sizing technique based on the ordering quantity being equal to the net requirements for a fixed time period.

Focus Forecasting A system that simulates the effectiveness of numerous forecasting techniques to determine the most effective one.

Forecast An estimate of future demand based on either qualitative judgement, quantitative computation of historical data, or a combination of both techniques.

FPO Firm Planned Order.

Gross Requirements The total of independent and dependent demand for a component not considering on-hand inventory and scheduled receipts.

Group Technology A system that identifies physical similarity of parts and provides both rapid retrieval of existing designs as well as facilitating cellular layout.

Horizontal Dependency The relationship of two or more components of the same parent.

Implode Tracing the usage from the bottom to the top of a bill of material using where-used logic.

Independent Demand Demand for an item that is unrelated to the demand for other items such as service part requirement.

Infinite Loading Calculation and scheduling of the operations required at work centers in the time periods required regardless of the capacity available to perform the work.

Input–Output Control A technique for capacity control where planned and actual inputs and planned and actual outputs of a work center are monitored.

Item Master File An item record containing descriptive data, control values, inventory status, requirements, and planned orders.

JIT Just-In-Time.

Job Shop A manufacturing organization in which the productive resources are organized according to function.

Just-In-Time (JIT) A philosophy for guiding manufacturing management with the goal of achieving maximum customer service while improving both quality and productivity. Emphasis is based on continuous improvement through the elimination of non-value-added activities.

Kanban A Just-In-Time pull system in which work centers signal with a card to withdraw parts from feeding operations. A second card may be used to authorize production.

Lead Time The span of time required to perform a process or a series of operations. The time span should include the time for the recognition of the need.

Least Total Cost A dynamic lot-sizing technique that calculates the order quantity by comparing setup and carrying costs and selects the lot where these costs are most nearly equal.

Least Unit Cost A dynamic lot-sizing technique that adds setup costs and carrying costs for each trial lot size, divides by the number of units in the lot size, and picks the lot size with the lowest unit cost.

Logistics The art and science of obtaining, producing, and distributing material and product in the proper place and in proper quantities.

Lot-for-Lot A lot-sizing technique that generates orders in quantities equal to the net requirements in each period.

Lot Size The amount of an item ordered for production or purchase.

MAD Mean Absolute Deviation.

Make-to-Order Product A product that is finished after receipt of a customer order.

Make-to-Stock Product A product shipped from finished goods "off the shelf" which has been finished prior to the customer order arrivals and based on forecasted demand.

Manufacturing Inventory Inventory required to meet production requirements as opposed to distribution inventory to meet customer demand.

Manufacturing Resource Planning (MRP II) A method for effective planning of all resources of a manufacturing company. It is an outgrowth and extension of closed-loop MRP with output integrated with financial reports.

Master Production Schedule (MPS) An anticipated build schedule that drives material requirements planning. It may be expressed in timing and quantities of end times, components, pseudonumbers, or planning bills of material.

Material Requirements Planning (MRP) A set of techniques that uses bill of material, inventory data, and the master production schedule to calculate material requirements. It makes recommendations based on time-phased net requirements of components.

Mean Absolute Deviation (MAD) The average of the absolute values of the deviations of actual values to expected values.

Mixed-Model Production Producing products in varying lot sizes with the goal to making in one day what is sold that day. This reduces finished goods inventory and assists in balancing component demand from feeding operations and suppliers.

Modular Bills of Material A type of planning bill arranged in product modules or options.

Move Card A Just-In-Time pull system; a signal authorizing the move from the outbound stock point of a feeding operation to the point of use.

Moving Average An arithmetic average of a specified number of recent observations. As a new observation is added, the oldest observation is dropped.

MPS Master Production Schedule.

MRP Material Requirement Planning.

MRP II Manufacturing Resource Planning.

Multilevel Bill of Material A display of all components at all levels directly or indirectly used in a parent, together with the quantity required of each component.

Net Change MRP An approach calling for a partial explosion whenever there is a change in requirements, open order inventory status, or bill of material. Only those parts affected by the change are recalculated.

Net Requirements The amount of material that needs to be ordered to cover the difference between on-hand and on-order status and gross requirements.

Non-Value-Added Activity Any activity that does not add value to a part in the transformation of raw material to the finished product.

Nonsignificant Part Numbers Part numbers assigned to each part but do not cover information about that part other than to identify.

Operation Sequencing A simulation technique for short-term planning of actual jobs to be run in each work center based on capacity and priority and then repeating the simulation for the next operation.

OPT Optimized Production Technology.

Optimized Production Technology (OPT) A scheduling philosophy centering on the management of bottlenecks through finite loading procedures.

PAB Projected Available Balance.

Parent The higher-level item in a bill of material with a component or components structured at the lower level.

Part Number A number that serves to uniquely identify a part.

Part Number Classification A method used in group technology to identify the physical or processing similarity of parts.

Part Period Balancing A dynamic lot-sizing technique that uses the least total cost method but adds a look-ahead/look-back routine before firming up the lot quantity.

Pegging The routine to identify for a given item the sources of its gross requirements and/or allocations.

Performance Measurements The tracking of actual results of selected activities and comparing them to goals or standards.

PFS Process Flow Scheduling.

Phantom Bill of Material A coding and structuring technique for subassemblies that permits MRP logic to drive requirements through the phantom item to its components but retains the ability to net against inventories of the item.

Planned Order A suggested order quantity, release date, and due date created by MRP processing based on net requirements. They exist within the computer and may change by the computer during subsequent MRP processing.

Planned Order Receipt The quantity planned to be received at a future date as a result of a planned order release.

Planned Shipment The same as an MRP planned order but used in DRP calculations.

Planned Shipment Receipt The same as an MRP planned order receipt but used in DRP calculations.

Planning Bill of Material An artificial grouping of items in a bill of material format to facilitate master scheduling.

Prioritizing Rules Differing priority decision rules for establishing the run sequence of orders. Which rule to use is the decision of the user and should be based on the existing manufacturing environments.

Process Flow Production Production with minimal interruptions in actual processing with queue time virtually eliminated.

Process Flow Scheduling (PFS) A method for planning equipment usage and material requirements that uses a process structure for scheduling calculations.

Process Manufacturing Production that adds value by mixing, separating, or forming in either a batch or continuous mode.

Product Structure The way materials go into the product during its manufacture.

Production Card In a Just-In-Time pull system, a signal used to authorize additional production.

Production Plan An agree-upon plan usually stated in monthly increments for production of product families and is the basis for the development of a more detailed master production schedule.

Projected Available Balance (PAB) The inventory balance projected out into the future considering on-hand inventory, requirements, scheduled receipts, and planned orders.

Projected On Hand Projected available balance excluding planned orders.

Pull System In-Just-In-Time material control, the withdrawal of inventory from an outbound stock point or supplier as demanded by signal from the using operations.

Purchase Order The purchaser's document used to formalize a purchase transaction with a supplier.

Push System In material control, the production of items or the issuing of material based on schedules planned in advance.

Queue The jobs at a given work center waiting to be processed.

RCCP Rough-Cut Capacity Planning.

Regenerative MRP An approach where the master production schedule is totally reexploded down through the bill of material with requirements and planned orders recalculated.

Regional Distribution Center A distribution center that services satellite distribution centers in nearby geographical areas as well as local customers.

Reorder Point A predetermined inventory level where if the stock on hand plus on order drops below that point, action is taken to replenish the stock.

Repetitive Manufacturing A form of manufacturing where various items with similar routings are made across the same process in repeating frequencies.

Resource Bill of Material See Bill of Resources.

Rough-Cut Capacity Planning (RCCP) The process of converting the production plan and/or the master production schedule into capacity needs of key resources in order to establish a feasible master production schedule.

Routing Details of manufacturing method of an item including operations, their sequence, work centers, and setup and run standards.

Safety Lead Time A time element added to normal lead time to advance the real need date to protect against fluctuations in demand and/or supply.

Safety Stock A quantity of stock planned to be in inventory to protect against fluctuations in demand and/or supply.

Scheduled Receipt Released purchased or manufacturing orders not yet received in stock.

Service Part Items planned to be used without modifications to replace an original part during maintenance.

Setup The work required of a machine, work center, or line from making the last good piece of an order to the first good piece of a new order.

Shop-Floor Control A system using shop-floor data to schedule, update status, and monitor material movement through the shop.

Significant Part Number Part numbers assigned to each part that identify the item and to convey certain information, such as the source of the part, the material in the part, the shape of the part, and so forth.

Single Source A supplier selected to have 100% of the business for a part, although alternate suppliers are available.

Single-Level Bill of Material A bill of material with all components shown as directly used in the parent item.

Single Minute Exchange of Die (SMED) A concept of setup time reduction.

SMED Single Minute Exchange of Die.

Sole Source The only supplier capable or available to meet the requirements of an item.

Standard Cost The target cost of an item including direct material, direct labor, and overhead charges.

Standard Deviation A measure of dispersion of data based on the deviation of actual values to expected values.

Subassembly An assembly that is used in a higher level to make up another assembly.

Supplier Certification A system to establish a supplier as a single source or the qualification at the part number level to replace incoming inspection with a sampling process.

Theory of Constraints (TOC) A management philosophy to identify core problems of an organization and the method to deal with the problems (constraints).

Time Fence A policy or guideline to note when various restrictions or changes in operating practices may or may not take place.

Time Phasing The technique of expressing future demand, supply, and inventories by time period.

Time-Phased Order Point (TPOP) A technique to control distribution or independent demand inventories using MRP logic. Time periods are used and therefore allow for lumpy demand.

TOC Theory of Constraints.

TPOP Time-Phased Order Point.

Tracking Signal A calculation based on deviations between a forecast and actual values to signal when the validity of the forecast may be in doubt.

Transient Bill of Material See Phantom Bill of Material.

Transportation Planning Report The summarized report of DRP planned shipments by ship date and distribution center.

Two-Bin System A fixed order system where an empty first bin calls for a replenishment order and the second bin, when full, will cover demand during lead time, plus safety stock.

Value-Added Activity An activity that adds value to a part in the transformation of raw material to the finished product.

Vertical Dependency The relationship of a parent and its component(s).

Wagner–Whitin Algorithm A mathematically complex dynamic lot-sizing technique that evaluates all options to arrive at the theoretically optimum ordering strategy for the entire net requirements schedule.

WIP Work in Process.

Work in Process (WIP) Products in various stages of completion throughout the plant from released raw materials through finished good at final inspection.

APPENDIX B
Certification Considerations

To prepare for APICS Certification, it is recommended that the candidate take the APICS sponsored 8-week classes or the 2-day review seminars. The material in this book will assist in the understanding of the principles and practices required for passing the Master Planning and the Material and Capacity Requirements Planning modules. The following body of knowledge information should be understood prior to taking the certification exams. The information is based on the contents in this book as well as certification course material.

MASTER PLANNING—FORECASTING

A forecast is an estimate of future demand that is of an uncertain nature, whereas order service is based on more certain aspects such as customer orders. The forecasting and order service activities combine together to make up the demand management function. Whereas demand for finished goods, service parts, and other independent demand items can and should be forecasted, dependent demand should be calculated rather than forecasted.

A forecast must not only state a projection of demand but must also project an estimate of the accuracy of the projection itself. An example would be that based on a forecast of 1000, the actual demand would be expected to be between 970 and 1030, 98% of the time.

The farther out into the future, the less accurate the forecast. Similarly, the more detailed the items to be forecasted, the less accurate the forecast. The forecast of a product group going out 2 months would be expected to be much more accurate than the forecast of each item in the product group with a 1-year horizon.

Forecasting methods can be as follows:

1. Qualitative, which is based on intuitive or judgmental evaluation and used for forecasting new products, processes, or markets. The qualitative method may also be used to adjust quantitative forecasts

2. Quantitative-intrinsic, which is based on a computational projection of historical patterns of the forecasted item's data
3. Quantitative-extrinsic, which is based on the computational projection of numerical relationships of external patterns

Weighted moving averages with the most recent periods receiving greater weight will respond more accurately to change than simple (unweighted) moving averages. Exponential smoothing is an easy way to calculate form of weighted moving averaging with the alpha (α) value defining the weighing factor. An α of 0.1 gives lower weight to the recent periods and can be used to smooth turbulent data. An α of 0.9 gives the recent data a heavy weight and forecasts the next period to almost duplicate the last observation.

The calculated standard deviation forecasts the variation (+ or −) to be expected from the forecast. The MAD (mean absolute deviation), which is easier to calculate than the standard deviation, will give a similar variation forecast.

A tracking signal will measure a bias or tendency of the actual observation to be either on one side or the other of the forecast. The demand filter is a calculation that will highlight any readings that are beyond reasonable expectations based on expected variances.

MASTER PLANNING—PRODUCTION AND RESOURCE PLANNING

Master planning includes the activities of production and resource planning as well as master scheduling. Production planning defines the anticipated work load required for product families, whereas resource planning defines facility and equipment requirements. The goal of production planning is to set the overall level of output and to integrate manufacturing with other business plan activities. Resource planning based on the production plan addresses resources which take long periods of time to acquire.

The functions of the production and resource planning process are as follows:

- Simulate alternative plans to achieve company goals
- Develop initial product group schedules
- Provide a time-phased production plan

The production plan should consider the costs of carrying inventory, personnel training, hiring and layoffs, overtime, subcontracting, and resource overuse.

The principles of production planning include the following:

- Inventories must be initially managed in the aggregate.
- General management is responsible for the preparation of the production plan.
- The master scheduler should be responsible for disaggragating the plan.
- The sum of the master production schedule requirements for all items should equal the production plan.
- A long-range production plan should be used to plan facility requirements.

The production plan states the volume loads by product family for monthly or quarterly periods. The resource plan identifies the facility and equipment required to support the production plan, whereas the master production schedule develops an anticipated build schedule, and the rough-cut capacity plan determines the resources required to meet the master production schedule.

Production planning is used for make-to-stock, make-to-order, and assemble-to-order environments. Planning can be to achieve level production rates (varying the inventory), level inventories (varying the production rate), or a compromise (varying both production rates and inventories).

Resource planning requires the identification of critical resources which then relate to the product structure through the bill of resources. The bill of resources lists the key resources needed to manufacture one unit of a selected product, family, or group. The key resources are defined by the user and can include money, capital equipment, labor, and/or material. The resource plan can then be calculated based on the production plan.

MASTER PLANNING—MASTER SCHEDULING

The master production schedule (MPS) is driven by the production plan, forecasts, customer orders, and projected inventories. The MPS drives the material requirements plan (MRP).

The items to be master scheduled are as follows:

- End items in a make-to-stock environment
- Components (including subassemblies) in an assemble-to-order environment
- Raw materials and components in a make-to-order environment
- Other user defined items such as critical impact, long lead time, or high-value items.

Planning bills are an artificial grouping of items in a bill of material format. They assist in more accurate forecasting, reducing the number of items to control, and in a more accurate planning process. Types of planning bills are as follows:

- Modular bills which are arranged by modules and/or options
- Common parts bills (or kits) which group components that are common for a number of individual products or product families into a single bill
- Super bills which tie together several modular bills and/or common parts bills at the highest level

The "projected available balance" identifies expected future inventory by time periods and considers both forecasts and customer orders. "Available-to-promise" is a technique which calculates inventory not committed to customer orders that will be available in certain time periods. Within the demand time fence, the forecast is consumed by customer orders. Outside of the demand time fence, the forecast is consumed by using the greater of customer orders or forecast.

When a work center's capacity is constrained (a bottleneck), the master scheduler may adjust the work center profile by the following:

1. Rescheduling
2. Increasing the work center capacity through actions such as overtime, adding shifts or personnel, or increasing outsourcing.

A firm planned order (FPO) is planned requirements with the quantity and due date fixed (or firmed). Firm planned orders are the normal method of stating the MPS and are useful in smoothing the schedule.

The final assembly schedule (FAS) is based on customer orders of finished end items. The FAS must be linked to and be compatible with the MPS.

MATERIAL AND CAPACITY REQUIREMENTS PLANNING—INPUTS AND OUTPUTS

The direct inputs to material requirements planning (MRP) are the master production schedule, the bill of material, and the inventory status. The direct outputs are a schedule of planned purchase order releases, a schedule of factory order releases, and recommended action notices. MRP is primarily an order scheduling system.

The direct inputs to capacity requirements planning (CRP) are planned factory order releases from MRP, open factory orders, routing, and work center data. The direct outputs are work center load reports and detailed factory order schedules.

The master production schedule (MPS) drives the MRP process, implements the production plan, and is the means to reconcile capacity with production to drive the MRP. The MPS planning horizon should be at least as long as the longest cumulative lead time for the items being produced. The planning horizon may have to be longer than the cumulative lead time in order to control capacity needs.

Whereas on-hand inventory balances, bill of material structures, and scheduled receipt data require a high degree of accuracy, planned lead times do not require as high a degree of accuracy but must be reasonable estimates. Stocks of inventory on hand that have been reserved for a specific use (allocated) must be recorded as such in the inventory status and reduced from the on-hand balance prior to the MRP run process.

A regeneration MRP system calls for a scheduled process run that is usually generated weekly. The system is therefore time-driven. All input data are loaded and a complete MRP plan is computed. All active item master records and bills of materials must be accessed and then processed level by level. The system will automatically flush out errors, but the plan is only current at the beginning of the first day.

With a transaction-driven net change system, the requirements and projected data are never erased. All transactions that affect inventory, bills of materials, or requirements information are processed, balancing records and reexploding whenever necessary to maintain an up-to-date plan. Only item records affected by transactions are rebalanced, and partial explosions occur when required to modify affected component records. Output reports are shorter but more frequent. Net change systems, due to more frequent processing, are more current than regeneration systems, but also cause a "nervous" output that can be difficult to manage. System nervousness may be addressed by postponing action on small requirement changes.

When problems arise at the component level that cannot be easily solved, a solution may be possible at a higher level. This is accomplished with bottom-up replanning using pegging data to review the higher-level parent requirements. Solutions may include compressing lead time, cutting order quantities, utilizing safety stock, substituting material, or, lastly, changing the master schedule.

Priority planning compares the due date of an open factory or purchase order with the need date of the latest MRP run. If the dates are different, the MRP output will call for a recommended action to expedite or deexpedite the order.

MATERIAL AND CAPACITY REQUIREMENTS PLANNING—THE MECHANICS

The conventional time assumptions in the MRP computations are the following:

- The gross requirements may be used at any time during the week.
- All scheduled and planned receipts are due at the beginning of the week.
- The projected on-hand inventory is as of the end of the week.
- All planned releases are released at the beginning of the week.
- Lead times are in weekly increments.

The projected on-hand balance = the projected on-hand balance from the previous period + scheduled receipts − gross requirements. If the projected on-hand is negative, it is a net requirement.

The planned order release of a parent multiplied by the usage of the component(s) equals the gross requirements of the component(s). The planned order release time is determined by backward scheduling the required due date receipt by the length of the lead time (lead-time offset). The quantity of the planned order is determined by lot size rules of the system.

Whereas scheduled receipts are considered in the projected on-hand balance calculation, planned receipts are not considered. Therefore, although planned receipts are calculated to cover net requirements, the projected on-hand will remain negative, increasing each time there is a net requirement. An alternative to the negative *projected on-hand* approach is the *projected available,* which treats planned receipts in the same manner as scheduled receipts. The projected available = the projected available from the previous period + scheduled receipts − gross requirement + planned receipts. The projected available will never be negative, as planned orders will be calculated to cover all requirements.

Allocated quantities are to be subtracted from the beginning on-hand balance or in the case of time-phased allocations from the appropriately time-projected on-hand balance. Although safety stock can be treated as an allocation, the certification exam assumes a second method which is to leave the safety stock in the on-hand balance but to have the system call for a planned order receipt when the projected on-hand or the projected available drops below the safety stock quantity (rather than zero).

A component may appear at more than one level in a product structure or structures used in an explosion. All gross requirements for a given component are accumulated for each level's requirements before balancing. This is accomplished through the use of a low-level code assigned to each part number. When there is independent demand for a component (such as for service usage), the forecasted usage for the demand should be entered as a gross requirement of the component.

A firm planned order (FPO) is a method for a planner to override the system and to freeze both the quantity and timing of a planned order. This technique can assist in the control of material or capacity problems. Pegging is a routine that allows the planner to trace the source of a lower-level requirement of a component by an upward level-by-level where-used review.

There are several lot sizing methods available to accommodate the demands of the system:

- Economic order quantity (EOQ). The lot size that balances order setup cost with inventory carrying cost and is optimum with even, regular, continuous usage.
- Fixed order quantity. Planner controlled quantity usually due to such factors as die life or purchase quantity discounts.
- Lot-for-Lot. A discrete quantity that covers the net requirements of each period. This method minimizes inventory but requires a low order setup cost.
- Least unit cost (LUC), least total cost (LTC), part period balancing (PPB), and the Wagner–Whitin algorithm. These techniques take into account the order setup cost and inventory carrying cost and then, through differing logic, allow for the anticipated nonuniform rate of requirements.

MATERIAL AND CAPACITY REQUIREMENTS PLANNING—CAPACITY REQUIREMENTS PLANNING

Capacity requirements planning (CRP) is used to validate the MRP plan by comparing projected work center loads with work center capacities. CRP is also the system that schedules planned orders at the work center level. The primary CRP application is for job-shop operations.

The rated or calculated work center capacity is based on adjusting theoretical capacity by utilization and efficiency factors. The capacity will normally be stated in standard machine hours for machining work centers and standard labor hours for assembly operations. A four-machine work center scheduled for two 8-hour shifts would have a weekly theoretical capacity of $4 \times 2 \times 8 \times 5 = 320$ scheduled hours. If the work center machine utilization factor was 0.80 (based on past experience) and the operation efficiency factor (a comparison of standard hours earned to actual charged) was 1.15%, the calculated weekly work center capacity would be $320 \times 0.80 \times 1.15 = 294$ standard hours.

The work center load is based on the open order status file as well as the MRP planned order releases. The work center profile is based on both the open and planned orders. The input to planned order scheduling is the routing data for the part number and work center data for the scheduled work center. Routing data elements include operation number, operation description, planned work center, standard setup time, and standard run times. The work center data elements are the capacity factors listed above as well as a queue allowance estimated for individual work centers.

Scheduling planned orders considers the following lead-time elements for each operation:

- Queue—Time in line waiting for the operation
- Setup—Time preparing for the operation
- Run—Time being worked on
- Wait—Time waiting to be moved
- Move—Time being moved to next location

In a job-shop operation, the queue time normally consumes the most time (in some cases up to 90–95% of the total time). Actual operation times are setup and run and are calculated in the work center load. The interoperation times, wait, move, and queue are not part of work center loads but are a big part of the lead time and, therefore, the work-in-process.

Items can either be forward or backward scheduled. Backward scheduling, which is most often used, starts with the order due date and, using scheduling data and system logic, back schedules to the order start date. Forward scheduling starts with the MRP start date and with similar standard data and logic forward schedules to an order completion date.

Infinite loading calculates the work center load by time periods based on part number and MRP needs, and without regard to capacity. Once calculated, the loads are compared to capacity, and if problems are indicated, balancing alternative must be addressed. Finite loading systems will not permit a work center to be overloaded but will move the overload to the next period. Finite loading or scheduling is usually addressed in the execution mode rather than planning. CRP will normally load without regard to capacity.

When infinite loading indicates capacity problems of either overloads or underloads, alterations to capacity or work loads should be made. Capacity may be increased by adding shifts, people, or equipment, or scheduling overtime. Capacity may be reduced through reducing shifts, hours, or people. The work load may be reduced by outside subcontracting, reducing lot sizes, or reducing the MPS (the least desirable). The work load may be increased by early order release, increasing the lot sizes, or increasing the MRP. The work load may be redistributed by the use of alternate work centers or routing, or revising the MPS.

APPENDIX C
Answers

CHAPTER 1

1. Modern methods of inventory management were made possible with the introduction of
   I. Kardex files
   II. machine tools
   III. computers
   IV. statistics

   a. I
   b. II
   c. III
   d. IV

c. Although the logic was understood, computers were required to handle the large amount of data.

2. Service level is
   a. part of the bill of material
   b. a measure of delivery performance
   c. unique part number
   d. a criteria of MRP

b. Service level is often expressed as shipped on time or percent of total number of items shipped.

3. Customer service is only measured by on-time delivery.
   a. True
   b. False

b. False. Customer service also includes other needs such as quality, reliable communications, and flexibility.

4. MRPII refers to
   a. material requirements planning
   b. manufacturing resource planning

c. closed-loop MRP
d. time-phased order point

b. Manufacturing resource planning

5. The economic lot size formula (EOQ) balances the cost of carrying the inventory with
   I. the reorder point
   II. the cost of setting up the run
   III. the selling price of the product
   IV. the lead time

   a. I and IV
   b. II
   c. I and III
   d. II and III

b. II—The balance is between carrying cost and setup cost.

6. MRP is designed for dealing with
   I. dependent demand
   II. independent demand
   III. discontinuous service
   IV. nonuniform demand

   a. I and II
   b. II and IV
   c. I, III, and IV
   d. II, III, and IV

c. I, III, and IV—All but independent demand can be based on MRP logic.

7. Which of the following applications would be best suited to an order point system?
   a. A product manufactured to customer order
   b. A subassembly that is used in one finished-goods item
   c. Component part inventory having a lumpy demand pattern
   d. A finished-goods item sold off the shelf with level usage

d. The order point system is best used for forecasted level usage.

8. MRP is a system for
   I. material requirements
   II. rescheduling recommendations
   III. execution

   a. I
   b. I and II
   c. I and III
   d. I, II, and III

b. I and II—MRP is for planning, not execution.

9. MRP planning is required for
   I. job-shop operations
   II. Just-In-Time facilities
   III. operations with bottlenecks
   IV. nonsynchronized flow operations

   a. I and II
   b. I and III
   c. II and IV
   d. All of the above

d. All of the above required MRP planning.

10. Implementation of an MRP system guarantees a successful manufacturing operation.
    a. True
    b. False

b. False—Not only must the MRP system be properly implemented, but the plans must then be executed according to plan.

CHAPTER 2

1. A forecasting system is required for which of the following elements of total demand of an item?
   I. Independent demand
   II. Dependent demand
   III. Replacement-part demand

   a. II
   b. I and III
   c. II and III
   d. I, II, and III

b. I and III—Dependent demand is based on the requirements of parent items rather than forecasts.

2. Ideally all manufacturing inventory would be
   I. considered finished goods
   II. consumed upon completion
   III. forecasted

   a. I
   b. II
   c. I and II
   d. All of the above

b. II—The goal is to minimize work-in-process investment.

3. If a forecast is 50/week, the lead time is 8 weeks, the safety stock is 4 weeks, and the lot size is 800, the reorder point is
   a. 400
   b. 600
   c. 800
   d. 1000

b. 600. (8-week lead time × 50) + (4-week safety stock × 50) = 600. The lot size is not a consideration.

4. The forecast is taken into account in
   I. distribution inventory
   II. MRP
   III. the master schedule

   a. I
   b. I and II
   c. I and III
   d. I, II, and III

c. I and III—MRP is driven by the master schedule which has taken the forecast into account. An exception would be when service parts usages are forecasted and are in the gross requirements of a component.

5. In bicycle manufacturing and distribution, the demand for a tire sold for replacement as well as used in assembly would be considered
   a. independent
   b. dependent
   c. mixed
   d. none of the above

c. Mixed—Dependent on the bicycles manufactured and independent based on replacement demand.

6. The master schedule takes into account
   I. the forecast
   II. finished goods inventory
   III. order backlog

   a. I
   b. I and II
   c. I and III
   d. I, II, and III

d. I, II, and III—All must be considered in calculating an anticipated build plan.

7. Distribution requirements planning (DRP) is used for
   a. manufacturing items with dependent demand
   b. purchased components
   c. relating branch warehouse demand to the manufacturing facility

c. Relating branch warehouse demand to the manufacturing facility. Manufacturing items with dependent demand would be MRP controlled as would purchased components.

8. If 80 are required and there are 50 in stock, 30 is the
   a. gross requirement
   b. available inventory
   c. net requirement

c. Net requirement; 80 is the gross requirement and 50 is the available inventory.

9. A time-phased order point will be used for
   I. finished goods
   II. a uniform continuous usage rate
   III. a nonuniform usage rate
   IV. dependent demand items

   a. I and II
   b. I and III
   c. I and IV
   d. IV

b. I and III—If the rate is continuous and uniform, a reorder point is used. Dependent demand items use MRP.

10. Safety stock may be based on
   I. a predetermined quantity
   II. anticipated usage for a given time period

   a. I
   b. II
   c. Either I or II
   d. Neither I nor II

c. Either method may be used depending on the situation.

CHAPTER 3

1. The bill of material is primarily a document for design engineering.
   a. True
   b. False

b. False—Design engineering will generate the bill of material, but manufacturing is the user.

2. The bill of material is used for
   I. MRP
   II. product costing
   III. defining the product

   a. I and II
   b. I and III
   c. II and III
   d. All of the above

d. All of the above

3. A separate part number must be generated when
   I. the item has changed identities
   II. the item is a parent
   III. the item is a component of more than one parent

   a. I
   b. I and II
   c. I and III
   d. All of the above

a. When an item requires a specific identity. The fact that an item is a parent or is used on more than one item does not require an additional number.

4. A transient subassembly or phantom will be
   I. coded as such
   II. have a lead time of "0"
   III. have a "lot-for-lot" lot size

   a. I
   b. I and II
   c. I and III
   d. All of the above

d. All of the features are required to allow the requirements to blow through the MRP calculation.

5. If a subassembly is immediately consumed in the assembly operation but is carried in stock for service usage, that subassembly can be
   a. modularized
   b. coded as a phantom
   c. eliminated from the bill
   d. purchased

b. Although not needed for assembly requirements, the subassembly must be maintained for service.

6. A finished item such as a bolt which is manufactured directly from purchased steel rod would be defined by a
   a. phantom bill of material
   b. planning bill of material
   c. single-level bill of material
   d. multilevel product bill

c. A single-level bill will define the total relationship.

7. The details of how to manufacture an item are found in
   a. the item master file
   b. the bill of material file
   c. the planning bills
   d. the routing file

d. The routing file lists operations, run and setup times, departments, and so forth.

8. The bill of material can be restructured for purposes of forecasting.
   a. True
   b. False

a. True—Restructured bills are called planning bills.

9. Planning bills assist in
   I. ease of forecasting
   II. master scheduling
   III. costing out finished goods

   a. I and II
   b. I and III
   c. II and III
   d. All of the above

a. I and II—Costing of the finished goods will require a product bill.

10. The bill of material is used for
   I. MRP
   II. defining the process
   III. cycle counting

   a. I only
   b. I and III
   c. II and III
   d. All of the above

a. I only—Cycle counting is item by item and does not consider bills of material. The routing file defines the process.

CHAPTER 4

1. The functions of the MPS over the short horizon are
   I. driving the MRP
   II. planning short-term capacity requirements
   III. recommending production of components
   IV. planning order priorities

   a. I, II, and III
   b. I, III, and IV
   c. I, II, and IV
   d. All of the above

d. All of the above.

2. The MPS should strive to
   I. maintain a balance between scheduled load and available productive capacity over the short term
   II. form basis for establishing planned capacity over the long term

   a. I
   b. II
   c. I and II
   d. Neither I nor II

c. Both I and II.

3. Development and administration of the MPS is the responsibility of
   I. marketing
   II. manufacturing
   III. finance

   a . I and II
   b. I and III
   c. II and III
   d. All of the above

d. All of the above.

4. The MPS must maintain MRP realism between planning and
   a. engineering
   b. finance
   c. execution
   d. marketing

c. Execution—The MRP assumes infinite capacity. The MPS is the control mechanism.

5. The master production schedule can supply the marketing department with information on the status of shipable end items for all of the following products EXCEPT

a. dependent demand components
b. single-model appliances
c. consumer package goods
d. custom-built machines

a. Dependent demand components are controlled through the MRP, not the MPS.

6. Exponential smoothing is a routine method for calculating
   a. safety stock
   b. forecasts
   c. standard deviations
   d. tracking signals

b. Forecasts—It is a method of using weighted averages.

7. Forecast bias is measured with
   a. exponential smoothing
   b. demand filters
   c. standard deviations
   d. tracking signals

d. Tracking signals will track if actual results show a bias on either side of the forecast.

8. A greater degree of accuracy will be possible if the forecast is
   I. by product groups
   II. for a short term

   a. I
   b. II
   c. I and II
   d. Neither I nor II

c. Groups measured over a short term are more accurately forecasted.

9. Forecast error can be measured through
   I. standard deviations
   II. mean absolute deviations

   a. I
   b. II
   c. I and II
   d. Neither I nor II

c. Both standard deviations and mean absolute deviations are based on deviations from forecast.

10. Demand data entry error may be detected by
    a. a tracking signal
    b. a demand filter
    c. second-order smoothing
    d. a weighted average

b. A demand filter will isolate readings that are beyond reasonable deviations.

CHAPTER 5

1. MRP evolved from combining the following two principles:
   I. Dependent demand calculation
   II. Independent demand calculation
   III. Time phasing
   IV. Inventory accuracy

   a. I and II
   b. I and III
   c. II and III
   d. II and IV

b. I and III—Time phasing and dependent demand are the principles of MRP logic.

2. The time-phased order point
   I. is a technique for controlling independent demand items
   II. has system processing logic similar to MRP
   III. is suited for service parts
   IV. ignores the aspect of specific timing

   a. II
   b. I and II
   c. I, II, and III
   d. All of the above

c. I, II, and III—Time-phased order points are based on specific timing.

3. Prerequisites for an MRP system are
   I. master production schedule
   II. bill of material
   III. unique part numbers
   IV. available inventory numbers

   a. I, II, and III
   b. I, II, and IV
   c. II, III, and IV
   d. All of the above

d. All of the above are prerequisites.

4. An MRP system answers the following:
   I. What material is needed?
   II. In what quantities?
   III. When is it needed?

   a. I and II
   b. I and III
   c. II and III
   d. All of the above

d. All of the above.

5. "Netting" consists of allocating against "gross"
   I. on-hand inventory
   II. on order

   a. I
   b. II
   c. I and II
   d. Neither I nor II

c. Both on hand and on order are considered available and will not require ordering.

6. The planning horizon
   I. is related to the longest cumulative procurement and manufacturing lead times for components
   II. can be less than the longest cumulative lead time

   a. I
   b. II
   c. I and II
   d. Neither I nor II

a. I—The horizon must cover the longest possible lead-time consideration.

7. A numbered-day shop calendar
   I. considers only scheduled working days
   II. makes scheduling arithmetic straightforward

   a. I
   d. II
   c. I and II
   d. Neither I nor II

c. I and II—Considers and calculates based on working days only.

8. The lead time of a manufactured item includes
   I. run time
   II. setup time
   III. queue time
   IV. wait time

a. I and II
b. I and III
c. I, II, and IV
d. All of the above

d. All of the above.

9. Regenerative MRP requires
   I. every MPS item must be exploded
   II. every active bill of material must be retrieved
   III. the status of every active item must be recomputed
   IV. voluminous output may be generated

a. I, II, and III
b. I, II, and IV
c. II, III, and IV
d. All of the above

d. All of the above.

10. The main difference between regenerative and net change MRP is
   I. The frequency of replanning
   II. what sets off the replanning process

a. I
b. II
c. I and II
d. Neither I nor II

c. Both I and II—Regenerative MRP is based on planned generating frequencies while net change is based on item transactions. In practice, the frequency could be the same for either method, but regenerative is usually run less frequently, such as weekly, compared to net change which might run daily.

CHAPTER 6

1. A firm planned order "freezes" the
   I. quantity
   II. timing

a. I
b. II
c. I and II
d. Neither I nor II

c. Both the timing and quantity are frozen.

2. If the setup time is decreased and the economic order quantity is recalculated, the quantity will

a. increase
b. decrease
c. remain the same

b. With less setup time, a smaller lot size is expected.

3. The fixed order quantity technique will vary
   a. the order interval
   b. the quantity
   c. quantity and interval
   d. neither quantity nor interval

a. The quantity will remain fixed, but the interval can vary.

4. If you wish to minimize inventory, the lot size technique that will be most effective is lot for lot.
   a. True
   b. False

a. True—With lot for lot, only what is needed is called for.

5. The EOQ calculation allows intermittent demand and does not require a steady demand rate.
   a. True
   b. False

b. False—The EOQ calculation assumes a steady demand rate.

6. Before rescheduling a manufacturing order, the planner should consider
   I. safety stock
   II. the parents' requirements
   III. capacity

   a. I
   b. I and III
   c. II and III
   d. All of the above

d. All of the above.

7. The following factors affect the computation of requirements:
   I. Product structure
   II. Lot sizing
   III. Lead times
   IV. Timing of end item requirements

a. I and II
b. I and III
c. I, II, and III
d. All of the above

d. All of the above.

8. To determine the effect a component change has on a parent, you would use
   a. MRP
   b. capacity planning
   c. pegging
   d. shop-floor control

c. Pegging—A specific where-used inquiry.

9. An allocated quantity in the inventory file indicates that the quantity is earmarked for
   a. a parent order
   b. a cycle count
   c. purging
   d. review

a. The quantity that will be required for a released parent order.

10. A planned order is the same as a scheduled receipt.
    a. True
    b. False

b. False—A planned order has not been released. A scheduled receipt is based on a released order.

CHAPTER 7

1. MRP systems assume that capacity considerations are in
   I. the master schedule
   II. the inventory files
   III. MRP logic
   IV. the bills of material

   a. I
   b. II
   c. I and III
   d. II and IV

a. The master schedule must consider capacity.

2. A material requirements planning system is designed to answer
   I. what can be produced with a given capacity
   II. what production is required to meet a given master production schedule

   a. I
   b. II
   c. I and II
   d. None of the above

b. II—MRP does not consider capacity. MRP II—manufacturing resource planning—does consider capacity.

3. Rough-cut capacity planning considers
   I. every work center
   II. key resources
   III. financial projections only

   a. I
   b. II
   c. I and II
   d. All of the above

b. Key resources are considered in rough cut capacity analysis.

4. The resource bill of material structure is based on
   a. single-level parts bill only
   b. multilevel parts bill only
   c. purchased parts
   d. whatever you want

d. Whatever you want and consider key resources.

5. A proper load projection has the following:
   I. Completeness
   II. Based on valid priorities
   III. Future visibility
   IV. Includes planned and open orders

   a. I, II, and III
   b. I, II, and IV
   c. I, III, and IV
   d. All of the above

d. All of the above.

6. The capacity of a work center is based on
   I. hours available
   II. machine utilization
   III. open orders
   IV. machine efficiency

a. I and II
b. I and IV
c. I, II, and IV
d. All of the above

c. I, II, and IV—Open orders reflect the work load, not the capacity.

7. The following are considered part of the work load:
   I. Queue time
   II. Run time
   III. Setup time
   IV. Wait time

   a. I and II
   b. II and III
   c. I and III
   d. All of the above

b. Run and setup time are part of the load calculation. All of the above are part of the lead time calculation.

8. Random variation of work arrival at a work center can cause
   I. a build up of queue
   II. machine down time due to lack of work

   a. I
   b. II
   c. I or II
   d. Neither I nor II

c. Either a buildup of work or a starved work center can be the effect of random variation.

9. In the lead-time calculation, random variation of work is compensated for by
   I. increasing the run estimate
   II. queue time allowance
   III. reducing the efficiency factor

   a. I
   b. II
   c. III
   d. All of the above

b. The queue time allowance.

10. The activities of a work center are monitored through
    a. MRP
    b. the master schedule
    c. rough-cut capacity planning
    d. input–output control

d. Input–output control

CHAPTER 8

1. If all distribution centers have level demand, the demand on the central supply center will also be level.
   a. True
   b. False

b. False—The sum of level demands are not necessarily level.

2. A forecast of a DC is considered a
   a. scheduled receipt
   b. planned order
   c. shipment release
   d. gross requirements

d. The gross requirement of the DC.

3. If the master schedule cannot meet the requirements of a DC, the remedy may be to
   I. reduce lot size of the order
   II. increase the safety stock
   III. use safety stock at the DC

   a. I
   b. I and II
   c. I and III
   d. I, II, and III

c. I and III—Do not attempt to increase safety stock if the MPS is already overloaded.

4. The distribution lead time includes
   I. order release and pick
   II. loading time
   III. unloading time
   IV. in-transit time

   a. I and II
   b. I, II, and IV
   c. I, III, and IV
   d. All of the above

d. All of the above.

5. To trace specific MPS demand back to distribution centers, you would use
   a. firm planned orders
   b. capacity planning

c. pegging
d. projected on hand

c. Pegging—The same approach as with MRP.

6. If the shipping schedule conflicts with DRP requirements, you should
   a. ignore the DRP
   b. change the shipping schedule
   c. move the planned order in closer to meet the shipping schedule
   d. move the planned order farther out to meet the shipping schedule

c. Move in closer, if possible, to meet the DC requirements.

7. The distribution demands of the central supply master schedule is the sum of the __________ of the DC centers.
   a. gross requirements
   b. planned order receipts at the DC
   c. planned order shipments to the DC
   d. safety stocks at the DC

c. The planned order shipments to the DC centers.

8. In a distribution network, safety stock located in a central supply location will cause __________ inventory than if stored at the DC warehouses.
   a. less
   b. more
   c. the same

a. Less—Due to positive and negative errors canceling out each other.

9. A planned inventory build up can be stored at
   I. the central supply location
   II. the DC warehouses
   III. an outside warehouse

   a. I
   b. II
   c. I and II
   d. All of the above

d. All of the above—Depends on specific situations.

10. A distribution requirements planning system differs from a material requirements planning system in that a DRP system
    a. performs time-phased, level-by-level netting

b. is primarily concerned with finished goods inventories
c. can consider existing scheduled receipts
d. can calculate and peg dependent demand

b. DRP is primarily concerned with finished goods, whereas MRP is concerned with dependent demands of components.

CHAPTER 9

1. The assumption that a shop order will be completed by the MRP generated due date always requires
   I. capacity planning
   II. finite loading
   III. infinite loading
   IV. operation sequencing

   a. I
   b. I and II
   c. I and III
   d. IV

a. I—Capacity planning is always required.

2. Multiple jobs may be scheduled at the same time when __________ is used.
   a. finite loading
   b. infinite scheduling

b. Infinite scheduling.

3. Finite scheduling assumes infinite capacity.
   a. True
   b. False

b. False—It loads to the capacity limitation.

4. Bottleneck scheduling assumes
   I. one or more bottlenecks exists
   II. all operations in use be finite loaded

   a. I
   b. II
   c. I and II
   d. Neither I nor II

a. I—Only bottleneck operations must be finite loaded.

5. Operation sequencing
   I. is a short-term planning technique
   II. looks ahead to the next operation

   a. I
   b. II
   c. I and II
   d. Neither I nor II

c. I and II.

6. Finite scheduling
   I. considers capacity at the work center
   II. loads only to capacity levels
   III. prioritizes the work load

   a. I
   b. II
   c. III
   d. All of the above

d. All of the above.

7. With infinite loading, you
   I. assume available capacity
   II. risk multiple jobs scheduled at the same time
   III. load only to capacity levels

   a. I
   b. I and II
   c. I and III
   d. All of the above

b. I and II—Infinite loading does not consider capacity levels.

8. The basic control document for job shop work-in-process is the
   a. labor ticket
   b. work order
   c. material requisition
   d. drawing

b. The work order.

9. Operation scheduling for the dispatch list can be calculated by either forward or backward scheduling.
   a. True
   b. False

a. True—The method to be used is dependent on the nature of the operation.

10. If the rate of unforeseen events are excessive, you must use
    a. operation sequencing
    b. finite loading
    c. infinite scheduling
    d. safety stock

d. Safety stock—Excessive unforeseen events cannot be scheduled.

CHAPTER 10

1. Process flow scheduling is based on bill of material structuring.
   a. True
   b. False

b. False—It is based on process routing.

2. MRP can be utilized in a continuous-process industry environment to
   a. generate shop orders
   b. reschedule production
   c. control raw material purchases
   d. expedite

c. Control of raw materials—The other three choices do not lend themselves to continuous processing.

3. Process manufacturing operations may consist of
   I. mixing
   II. forming
   III. separating
   IV. job lot machining

   a. I
   b. II and III
   c. I, II, and III
   d. All of the above

c. I, II, and III—Job lot machining would be in a job-shop manufacturing environment.

4. Fabrication and assembly operations may be accomplished with
   I. process manufacturing
   II. job-shop manufacturing
   III. repetitive manufacturing

   a. I
   b. I and III
   c. II and III
   d. All of the above

c. Either job shop or repetitive, depending on the routing and repetitive nature of production.

5. The bill of material in a process manufacturing environment will tend to be shallow.
   a. True
   b. False

a. True—There will be no or few subassembly levels.

6. Process manufacturing can be in a ________ mode.
   I. continuous
   II. batch

   a. I
   b. II
   c. I and II
   d. Neither I nor II

c. Either continuous or batch modes can be used in process manufacturing.

7. In process manufacturing, the flexibility of the MPS is dependent on
   I. available raw material
   II. capacity

   a. I
   b. II
   c. I and II
   d. Neither I nor II

c. Both material and capacity must be considered.

8. In process manufacturing, if lot size constraints require batch controls for WIP, ________ is recommended.
   a. reorder points
   b. rough-cut capacity planning
   c. MRP
   d. infinite scheduling

c. MRP controls stocking levels in the process.

9. With continuous processing, a high degree of WIP shop floor control is required.
   a. True
   b. False

b. False—Continuous flow calls for less shop-floor control.

10. The goal of repetitive manufacturing is the same as the process manufacturing's goal.
    a. True
    b. False

a. True—Continuous flow is the goal.

CHAPTER 11

1. Effective implementation of Just-In-Time should result in which of the following?
   I. Maximum work center utilization
   II. Employee satisfaction
   III. Development of detailed cost accounting data
   IV. Increased total business productivity

   a. I and III
   b. II and IV
   c. I, II, and IV
   d. All of the above

b. II and IV—Maximum work center utilization and detailed cost accounting data are not part of JIT philosophy.

2. Long-term agreements with suppliers may include which of the following types of special arrangements?
   I. Simplified paperwork systems
   II. Returnable containers
   III. Specially designed containers
   IV. Effective personal contacts at each site

   a. IV
   b. II and III
   c. I, II, and III
   d. All of the above

d. All of the above.

3. To which of the following types of delivery can kanban methods be applied?
   I. Within a plant
   II. From plant to plant
   III. From suppliers to plants

   a. I
   b. I and II
   c. II and III
   d. All of the above

d. All of the above.

4. JIT training should increase the capabilities of operations supervisors to do all of the following EXCEPT
   a. balance work load in cells
   b. plan area production
   c. keep the work force running its machines
   d. coordinate preventive maintenance

c. Machine utilization is not a major concern.

5. Which of the following will occur when MRP and JIT are integrated?
   a. MRP will continue to support both operations planning and control.
   b. MRP will continue to support operations planning and JIT pull systems will support operations control.
   c. MRP will continue to support operations control and JIT pull systems will support operations planning.
   d. JIT pull systems will eliminate the need for MRP.

b. MRP for planning, JIT to execute.

6. Small-lot production contributes to reduction of all of the following EXCEPT
   a. work-in-process inventory
   b. need for operator skills
   c. fluctuation of load on work centers
   d. the number of defective units

b. The operator skills cannot be reduced.

7. One of the objectives of JIT flow through production is to maximize machine utilization.
   a. True
   b. False

b. False.

8. In the implementation of JIT, which of the following is the most significant change for the accounting function?
   a. Cellular processes no longer report every operation to accounting.
   b. Budgeted scrap and rework costs that create variances are reduced.
   c. Physical inventory of work-in-process inventory is more time consuming.
   d. Costs of incentive programs increase.

a. Accounting data requirements are reduced.

9. Characteristics of parts containers appropriate for a JIT pull system include all of the following EXCEPT
   a. a design that allows their movement by hand without undue strain
   b. physical size appropriate for the consuming work center
   c. capacity appropriate for the lot size of the producing work center
   d. a means of protecting the parts from handling damage

c. The lot size may be in multiples of the containers.

10. Which of the following software capabilities is used so that transaction costs are minimized when JIT is implemented in an MRP environment?
    a. Pegging
    b. Back-flushing
    c. Forecasting
    d. Lot sizing

b. Back-flushing reduces the number of transactions.

CHAPTER 12

1. In the implementation of MRP, the leadership role should be
   a. production control
   b. top management
   c. data processing

b. Top management leadership is essential.

2. In MRP implementation, which of the following is an initial step?
   I. Project organization
   II. Education
   III. Performance goals
   IV. Software selection

   a. I　　　　c. II
   b. I and III　　　　d. IV

c. Education is required for making the decision to implement MRP.

3. MRP education is required at what organization levels?
   I. Top management
   II. Manager
   III. Supervisory and staff
   IV. Operator

a. I and II
b. II
c. II and III
d. All of the above

d. All of the above.

4. MRP implementation is the primary responsibility of operating people rather than data processing.
   a. True
   b. False

a. True.

5. Specific detailed implementation issues should be addressed by
   a. the steering committee
   b. the project team
   c. the task force

c. The task force is organized for specific issues.

6. The major system training emphasis should occur during
   a. justification
   b. initial MRP education
   c. the conference room pilot
   d. software selection

c. A major objective of the conference room pilot is training.

7. Performance measurements should be initiated
   a. prior to MRP implementation
   b. when the conference room pilot is underway
   c. during the initial cutover
   d. when implementation is complete

a. Prior to MRP implementation.

8. Data accuracy goals can be more easily attained once the MRP system is successfully implemented.
   a. True
   b. False

b. False—Data accuracy is required for a successful implementation.

9. Bridge programs are required for
   a. data accuracy
   b. software evaluation
   c. companywide goal establishment and monitoring
   d. data transfer from the old to new system

d. Data transfer from the old to new system.

10. During the initial implementation of a new MRP system, inventory levels may
    a. increase
    b. decrease
    c. remain the same

a. There may be a temporary increase due to the remnants of the old system.

Index